NEVADA BUREAU OF MINES AND GEOLOGY SPECIAL PUBLICATION 16

Geologic Tours in the Las Veg

Expanded Edition with GPS Coordinates

by
Joseph V. Tingley, Becky W. Purkey, Ernest M. Duebendorfer,
Eugene I. Smith, Jonathan G. Price, and Stephen B. Castor

illustrations:
Kris A. Pizarro

publication design:
Jack P. Hursh

2008

Bitter Spring Valley.

Photo: David Gray

Mackay School of Earth Sciences
College of Science
University of Nevada, Reno

How To Use The Trip Guides

Trips 1 through 4 are loops beginning and ending in Las Vegas. Trip 5 is a collection of points of interest in Las Vegas Valley that you can visit in any order. A brief description of the highlights of each trip and its total mileage (except for Trip 5) are given at the beginning of each trip log.

Throughout the log, we reference directions by use of numbers as they appear on the face of a clock: 9:00 is to the left, perpendicular to the direction of travel; 12:00 is straight ahead; and 3:00 is to the right, perpendicular to the direction of travel. Turnoffs to key roads, exhibits, etc., are noted at many mileage points. Do not turn off unless instructed to do so.

On the longer trips, you will be instructed to reset the odometer to 0.0 at some places along the route; this makes it easier for you to choose or omit portions of the trip and also reduces errors caused by variations in odometers.

Most of the longer text descriptions are not tied to specific mileage points and can be read while you are stopped at view points. For the log and text descriptions at the mileage points, however, it works best if a passenger does the reading so the driver can devote full attention to driving.

A table of Global Positioning System (GPS) readings has been included as an appendix to the 2008 revision of this book (see p. 141). Readings were taken at major points of each trip log and keyed to the cumulative mileage figure shown in the text for the point. A GPS point number has been inserted in the text, such as "(*GPS 21*)", and the latitude and longitude of the point are listed in the table by that number. The reading is given in the format N00 (degrees) 00.000 (minutes), W000 (degrees) 00.000 (minutes) (for example N36 02.517 W115 11.275). Also included in the table are brief comments intended to help locate the point.

Travel in the desert can be dangerous for the unprepared traveler. Always carry water, maps, clothing to suit the weather, tools for emergency car repair, and maybe some snacks in case you spend longer enjoying the scenery of this magnificent area than you originally planned.

Please help protect the fragile environment of the desert. Never drive off highway, and don't venture down any of the many unimproved roads you might see leading off into the desert or down washes.

Contents

Photo: Mark Vollm

Ancient Puebloan petroglyphs etched into Aztec Sandstone above Red Spring in the Calico Hills, Red Rock Canyon National Conservation Area. ▶

Acknowledgments for the Expanded Edition:

This edition of Geologic Tours in the Las Vegas Area draws heavily on material from the 1994 edition and the authors of that publication deserve special thanks for allowing their work to be built upon and expanded. We also wish to thank them for their consultation and advice on compiling this expanded edition. Jon Price, Director of the Nevada Bureau of Mines and Geology, agreed to the expansion of the publication, and let us publish it in the larger, spiral-bound format. Susan Tingley provided encouragement throughout the project, helped rerun and update the trip logs, and drafted the maps and sketches in Trip 5. It would not have been possible to publish this work without the efforts of Kris Pizarro and Jack Hursh. Kris prepared the maps and sketches for everything except Trip 5 (including redrafting all figures used from the first edition), wrote the material used in the plant and animal inset boxes, and prepared the photographs used throughout the text. Kris's touch has transformed many photos into exceptional presentations. Jack Hursh designed and typeset the book, and provided many of his own excellent photos. Dick Meeuwig edited both the text and figures and provided helpful suggestions in many places.

The Bureau of Land Management (BLM) has provided us with financial support for the printing of this book. Tom Leshendok, BLM State Office, recommended this project for funding, and we thank him and the management and staff of the BLM Las Vegas Field Office for this support. Tom also reviewed the manuscript, and his comments improved the final product.

We gratefully acknowledge the time and effort that our reviewers gave to us; they provided us with new insights into the subject matter, and the manuscript has been greatly improved and clarified by their work. Reviews were provided by Neil Brecheisen, BLM, Carson City; Rebecca Range, BLM, Las Vegas; and Katherine Rohde, National Park Service, Boulder City.

Many of the photos used are from the first edition. Several photographers deserve thanks and lots of credit for allowing us to use their photos in the expanded edition. We would not have been able to complete this work without their contributions. New photos used in the expanded edition have been contributed by Mark Vollmer, Jack Hursh, Patti Gray, David Gray, Carol McKim, Kris Pizarro, Steve Castor, Susan Tingley, Jim Faulds, Douglas Filer, Roy W. Cazier, Larry Anderson, and Claire Lanouette. Kate Sorom, BLM Red Rock Canyon National Conservation Area (RRCNCA), sent us historical photos of the Sandstone Quarry in Red Rock Canyon. The RRCNCA also supplied some of the wildlife photos. Ralph Bennett gave us pen and ink sketches that we have used in several places throughout the text, and we have also used sketches by Larry Jacox.

Finally, discussions about the complex geology of the Las Vegas area with NBMG geologists Jim Faulds, John Bell, and Kyle House have contributed a great deal to the geologic interpretations presented herein.

Joseph V. Tingley, Economic Geologist
Nevada Bureau of Mines and Geology
August 2001

Acknowledgments for the First Edition:

The authors are grateful to the individuals and organizations that contributed their varied talents and services to the preparation of this publication. Nevada Bureau of Mines and Geology staff include: Dick Meeuwig, who edited the manuscript; Kris Pizarro and Jan Walker, who produced the fine maps, geologic illustrations, and botanical drawings from the authors' sketchy rough drafts; Rayetta Buckley, who composed and typeset this publication; and John Bell and Don Helm, who provided information from their current research on the recent faulting and subsidence issues addressed in Trip 5.

Outside of the NBMG, Tom Purkey donated two weeks assisting in several field checks of every mileage point. Art Gallenson, pilot with Lake Mead Air in Boulder City, took one author on a smooth, but not uneventful, ride over the guidebook area to obtain aerial photographs. He graciously provided his camera when hers broke early in the flight. Stephen Rowland, Professor of Geology at the University of Nevada, Las Vegas kindly offered his support throughout the completion of this project and helped in providing slides and sketches for one of the trips.

Except as noted in the captions, all photographs were taken by the authors. Historical photographs were reproduced with permission from the Nevada Historical Society and the California Division of Mines and Geology.

We especially want to thank the knowledgeable individuals who took the time to read this document and make valuable suggestions that substantially improved the final product: Joseph V. Tingley, economic geologist with the Nevada Bureau of Mines and Geology; Leslie G. McMillion, consulting hydrologist in Las Vegas; and Maxine Shane, public affairs specialist with the Bureau of Land Management in Reno.

Without the generous financial support of the U.S. Department of the Interior, Bureau of Land Management, this guidebook would not have been published. We especially want to thank Larry Steward, Thomas Leshendok, and Neal Brecheisen of the Reno Office of the Bureau of Land Management for recommending this project for funding and helping expedite its completion.

Becky Weimer Purkey, Geologic Information Specialist
Nevada Bureau of Mines and Geology
Ernest M. Duebendorfer, Associate Professor of Geology
Northern Arizona University
Eugene I. Smith, Professor of Geology
University of Nevada, Las Vegas
Jonathan G. Price, Director/State Geologist
Nevada Bureau of Mines and Geology
Stephen B. Castor, Research Geologist
Nevada Bureau of Mines and Geology

January 1994

Artwork: Larry Jacox

INTRODUCTION

Las Vegas, set in the southern tip of Nevada, is surrounded by some of the most spectacular geology in the American Southwest. Looking outward from the glittering city center, one can see evidence of nearly two billion years of the Earth's history expressed in the rugged mountains that border Las Vegas Valley. Thick sections of limestone, dolomite, sandstone, and shale, similar to rocks found in the Grand Canyon to the east, can be seen in Frenchman Mountain, the Spring Mountains, Red Rock Canyon, and Valley of Fire. These rocks have been cut, pulled apart, and in some cases moved for many miles along faults that define this area as part of the Basin and Range province. There are remnants of volcanoes and the volcanic rocks that erupted from them, and there are deposits of young sedimentary rocks that fill basins formed by faulting in the last few million years. Erosion of these geologic features by wind and water in this desert climate has resulted in the spectacular landforms that can be seen today.

Each of the five geologic tours in this guidebook takes the traveler to a different area of special geologic interest within a short driving distance of Las Vegas. In addition to geology, the trip guides tell a little of the people who shaped the human history of this part of southern Nevada, such as prehistoric American Indian salt miners, Spanish explorers, mountain men, Mormon settlers, and gold miners. Touching on natural history, there are comments on some of the plants and animals that make this area more than just stunning desert landscape. Depending on your trip choice, you can visit a National Conservation Area, two National Recreation Areas, three Nevada State Parks, and even a National Wildlife Refuge catering to desert bighorn sheep. These trips provide options for side trips, hiking, camping, wildlife viewing, and even fishing and boating if you elect to explore Lake Mead.

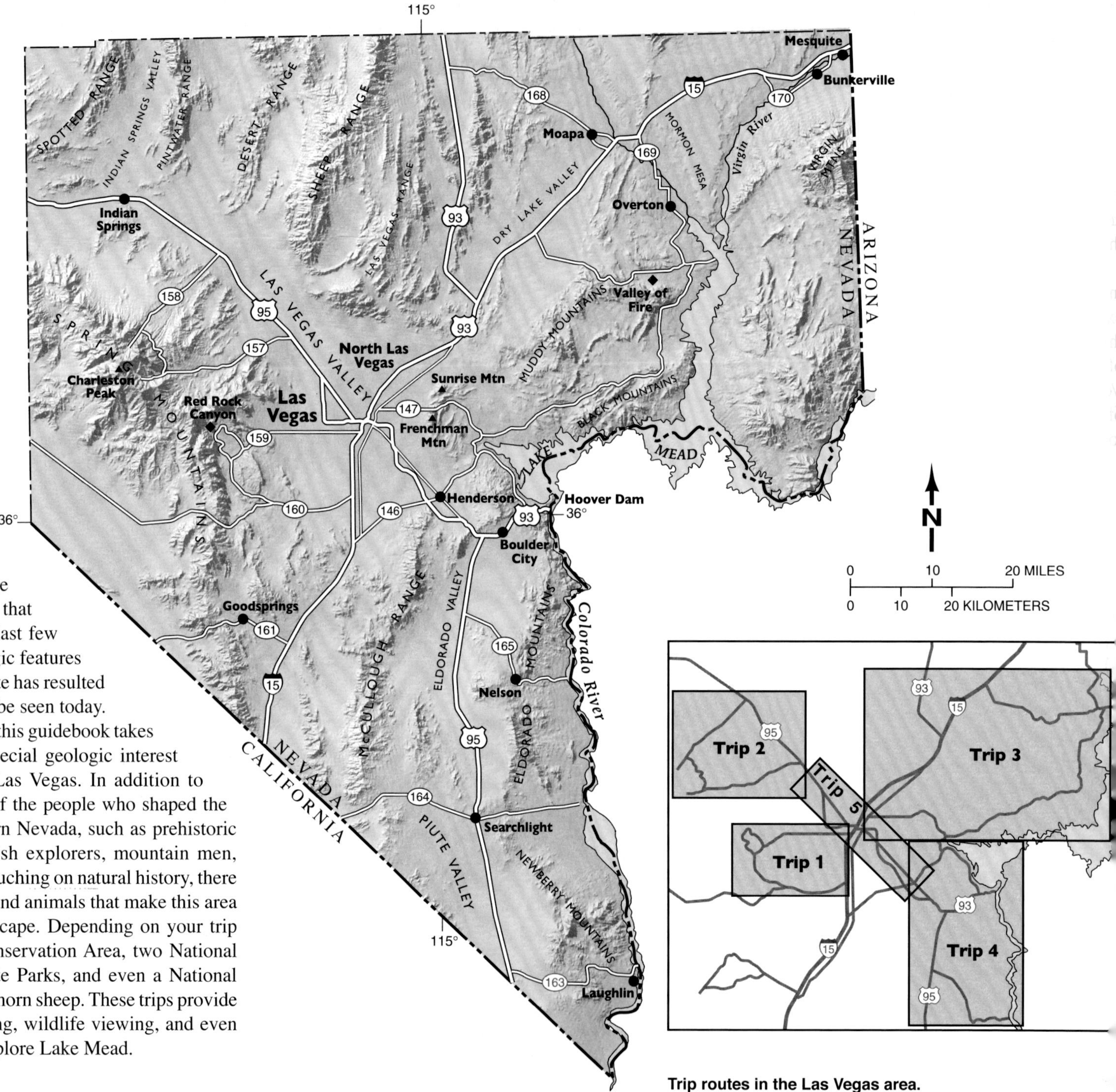

Trip routes in the Las Vegas area.

Physical Geography

The Las Vegas area is in the Basin and Range province, a region characterized by a series of generally north-trending mountain ranges and intervening valleys filled with eroded sediments. This province extends from southern Idaho through Nevada, Utah, Arizona, and New Mexico, into westernmost Texas and northern Mexico. The transition zone between the Basin and Range province on the west and the geologically stable Colorado Plateau province on the east begins just east of Las Vegas.

Most of the drainage in southern Nevada valleys is internal, that is, streams flow into valleys that have no external drainage. The Colorado River is an exception to this pattern. It drains an area along the southeastern margin of the state on its way to the sea. Elevations in the Las Vegas area range from a low of 450 feet along the valley of the Colorado River at the southernmost tip of the state to a high of 11,918 feet on Charleston Peak.

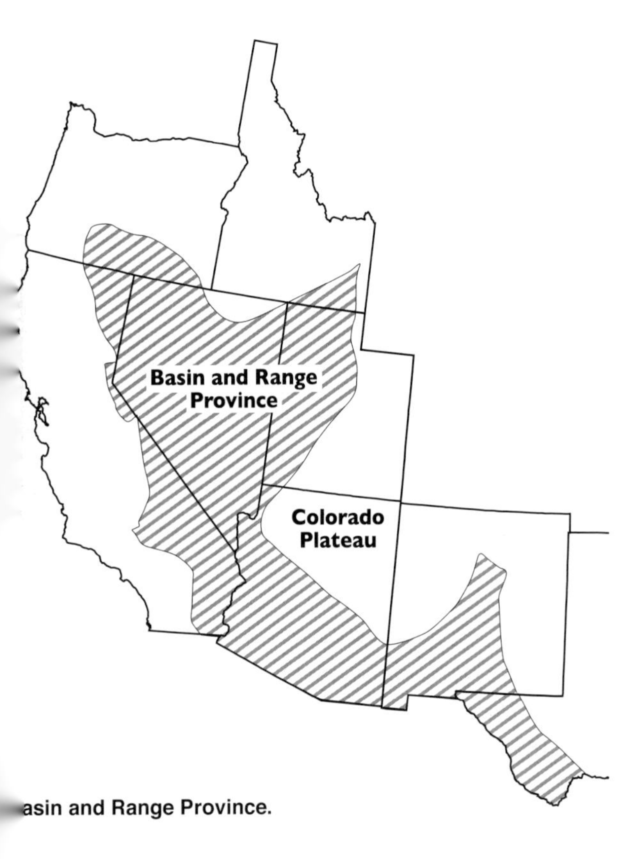

Basin and Range Province.

The climate ranges from hot and arid at the lowest elevations, to cooler, semiarid conditions in the higher basins, to more humid alpine climates in the high mountains. In most of the Las Vegas area, daytime summer temperatures can reach well over 100°F, but winter temperatures may plunge below freezing. Precipitation is less than 10 inches per year and occurs mainly from December to March from storms generated by Pacific Ocean currents. Summer storms may also provide measurable precipitation, sometimes in violent downpours. Rain shadow effects are evident on the leeward side of the higher ranges where intermontane basins are arid to semiarid. Evidence for a much wetter and cooler climate just a few thousand years ago can be seen on several of the trips.

Las Vegas Valley and most of the other valleys in the Las Vegas area are in the Mojave Desert, a transitional area between the Great Basin Desert to the north and the Sonoran Desert to the south. The Mojave Desert hosts a large variety of plants and animals that are well adapted to the hot, arid environment.

Plants along perennial streams and on playas located at or near the water table in the lowest parts of valleys are generally phreatophytes (plants that send down deep roots to obtain water from near or below the water table). In this area, these plants include tamarisk, mesquite, salt grass, greasewood, rabbitbrush, pickleweed, and iodine bush. Phreatophytes discharge large amounts of groundwater through transpiration. The alkali flats are home to mixtures of phreatophytes and non-phreatophytes (shallow-rooted plants that depend on near-surface water) such as shadscale, bursage, desert holly, creosote bush, Mormon tea, catclaw, alkali buckwheat, Mojave aster, and desert alyssum.

The lower slopes of alluvial fans that flank the mountain ranges, where the water table is considerably deeper, support communities of creosote bush, white bursage, catclaw, saltbush, little rabbitbrush, mallow, Mormon tea, sandpaper plant, brittlebush, littleleaf ratany, shadscale, fluff grass, and several species of cactus. Higher up on the alluvium and into the foothills, forests of Joshua trees and Spanish bayonet yucca are common, accompanied by blackbrush (the predominant small shrub) and various types of cactus and sagebrush. In riparian or marshy areas along streams at this elevation, reedbush, quailbush, mesquite, arrowweed, shadscale, wild grape, box thorn, Mormon tea, cattail, tamarisk, salt grass, bursage, indigo bush, desert willow, and various flowers are found.

Above 4,500 feet, wooded areas of juniper and piñon dominate. Above 7,000 feet in the Spring Mountains, ponderosa pine, Douglas fir, white fir, Englemann spruce, and some bristlecone pine are found. Oak and aspen also make up this forest environment along with small shrubs and grasses.

Insects, spiders, reptiles, and the many mammals, from kangaroo rats to desert bighorn sheep, are as varied as the plant life. While plants have adapted their leaves, needles, woody stems, germination cycles, and root systems to the desert, the animals have adapted their diet, water sources, mating habits, and body form and structure to thrive in this environment.

Photo: Mark Vollmer

Joshua trees along Kyle Canyon, Spring Mountains.

Photo: Mark Vollmer

Clearing winter storm over the Spring Mountains seen from S.R. 157 below Kyle Canyon.

Cultural History

Archaeological discoveries at such places as Tule Springs north of Las Vegas, Gypsum Cave near Frenchman Mountain, and the Lost City near Overton give evidence of early habitation of this area from the Desert Culture of 10,000 years ago to the end of the Pueblo Culture in 1150 A.D. Paiute people, who were living here when the first Europeans arrived on the scene in the 18th century, are thought to have entered the Las Vegas area around 1000 A.D. Traditional teachings of the Paiute people, however, tell that they have occupied these tribal lands since the beginning.

This area was within the vast territory claimed by Spain in the 17th century. The Spanish were navigating the Colorado River from the south as early as the 18th century and are credited as the first white men to find gold in the Eldorado Canyon area. Francisco Garcés, a Spanish priest, is thought to have passed through Las Vegas Valley in 1776. After the Mexican Revolution in 1821, this territory became part of the Republic of Mexico.

By the 1830s, early Mexican traders and immigrants on the Old Spanish Trail between Santa Fe and Los Angeles (which essentially follows the route of Interstate 15 through this area) were regularly using the springs at Las Vegas and in the Red Rock area. The Mexicans named the area Las Vegas, or The Meadows. Kit Carson and Jedediah Smith were also famous travelers through the area.

After the Mexican War in 1846–1848, the Old Spanish Trail was abandoned by Mexican traders. With the signing of the Treaty of Guadalupe Hidalgo in 1848, title to this territory was transferred to the United States. Mormon missionaries began using a trail between Salt Lake City and California. The church established a settlement, Mormon Fort, at the springs in Las Vegas in 1855, but soon abandoned it. Afterward, ranches were established around the springs; of note were the Stewart and Kyle (Kiel) ranches.

With the establishment of the San Pedro, Los Angeles and Salt Lake Railroad through Las Vegas in 1903, the Las Vegas townsite was established with a population soon totaling about 2,000. The railroad also stimulated mining in the nearby districts of Nelson, Searchlight, and Goodsprings.

During World War II, when Basic Magnesium established an important industrial area in Henderson, the population of Las Vegas rose to 17,000 people. Las Vegas Valley continues to grow, mainly as an entertainment mecca, and as of 2000 had a population of about 1.4 million.

Trails of early explorers and immigrants that crossed southern Nevada.

▲ **Stewart Ranch in the 1880s (site of Mormon Fort in 1855 to 1857 and Las Vegas Ranch 1867 to 1882).**

Photos: Nevada Historical Society

Fremont Street and railroad depot in Las Vegas in about 1910.

Artwork: Larry Jac

Geology

One of the most delightful aspects of the Las Vegas area is the desert scenery, particularly in the evenings and early mornings, when shadows are long and the landscape acquires contrast and provides a sense of serenity that is well-known to longtime area residents. This scenery—the mountains, canyons, valleys, alkali flats, and other features of the natural landscape—results directly from geologic processes of rock formation and erosion. In short, the foundation of desert scenery is geology.

Geology is also the dominant influence on other characteristics of the land, including water and soil distribution, that ultimately determine where and how plants and animals live. In addition, geology is an important factor in the human environment. Early human inhabitants lived where they could find water, food, and natural shelter. Most of the later human settlements, including Las Vegas itself, were built adjacent to springs or other water sources, or sprang up near valuable mineral deposits. Although we are not as dependent on the natural environment as our ancestors, geology continues to influence our lives.

To better understand the geology of the Las Vegas area as it unfolds along each of our trip routes, some basic principles are outlined in the following section. There is background on minerals, rocks, and geologic structure. We include some thoughts on plate tectonics, as the subject applies to southern Nevada, and we review the mineral deposits of the area. While reading these sections, it will be helpful to refer to the glossary at the back of the book for definitions of specific terms. For a more detailed look at the geology of Nevada and the Great Basin, refer to *Geology of Nevada* (Nevada Bureau of Mines and Geology Special Publication 4) by John Stewart, and *Geology of the Great Basin* by Bill Fiero (University of Nevada Press).

Photo: Jim Faulds

Alluvial fans contribute to the desert landscape throughout the southwest. This fan, formed at the mouth of a steep, rugged canyon on the northwest side of Sunrise Mountain, clearly shows why they are called "fans" (see sketch of desert landforms, page 105). The runways of Nellis Air Force Base are in the upper left of the photo.

Minerals and Rocks

Rocks are made of minerals. A mineral is a naturally occurring, inorganic, solid element or combination of elements, with a definite chemical composition and a regular internal crystal structure. That is why minerals have certain definitive shapes and physical properties.

Rocks are classified into three major types depending on how they were formed. **Igneous rocks** are those that formed from magma (molten rock) from deep below the Earth's surface. Igneous rocks are given names on the basis of their composition and mode of formation. Most rising magma never reaches the surface, but cools slowly at depth, forming masses of plutonic rock such as stocks, batholiths, and dikes. Granite is an example of plutonic rock that contains relatively large amounts of silica and alkali metal elements. Some magma does reach the surface of the Earth through openings in the crust, pouring out in calm flows or erupting violently and, in some cases, forming volcanoes. Magma that reaches the surface is called lava until it solidifies into volcanic rock. Basalt is a volcanic rock that contains relatively little silica and large amounts of iron and magnesium.

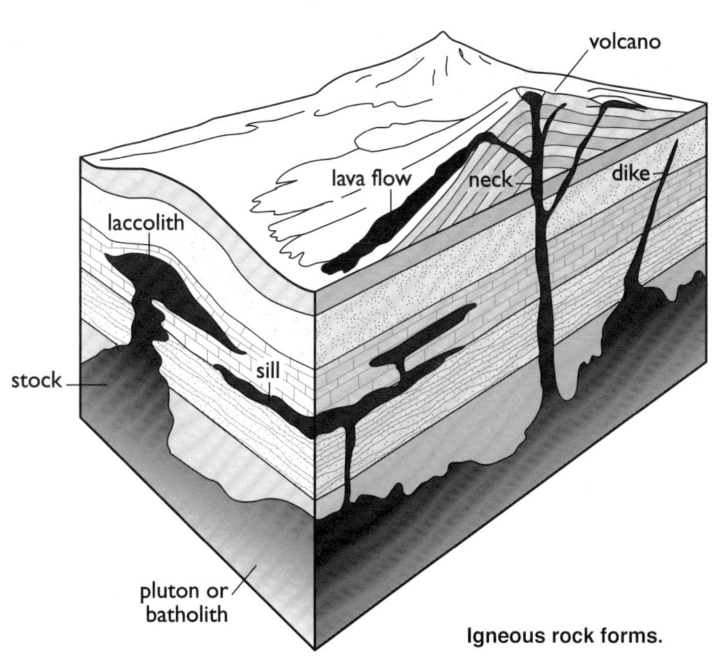

Igneous rock forms.

Sedimentary rocks are formed in two ways. Clastic sedimentary rock, such as sandstone, forms from accumulations of particles or detritus of preexisting rocks that were broken down and transported by the action of water, wind, or ice and deposited in a new location (in oceans, rivers, lakes, or desert basins). Nonclastic sedimentary rock is produced by chemically or biologically formed material precipitated out of certain bodies of water. Examples include limestone and evaporite rocks such as gypsum. In time, the deposited rock particles and precipitates harden to form solid rock.

Sedimentary rocks may contain fossilized remains of plants and animals, traces of their existence such as burrows or tracks, or preserved sedimentary structures such as cross beds in sandstone or mud cracks in siltstone that tell us of the life and terrestrial environments that existed while that rock was being deposited. Fossils, particularly those types that existed for a geologically short time, are especially useful in correlating rocks of the same age around the world.

Finally, **metamorphic rocks** are preexisting rocks that have been changed by heat, pressure, and chemically active fluids, generally while in the solid state. Metamorphic rocks are physically different from the original rocks from which they were formed. They are commonly denser and tougher, and the original mineral grains have undergone changes in size or shape, or entirely new minerals have been formed. If high temperatures were involved, the former rocks may have been partially melted. Gneiss and schist are two common types of metamorphic rock.

All three rock types are abundant in the Las Vegas area. They record the processes that have shaped the desert landscape.

Geologic Structures

Many of the geologic structures that can be examined in the Las Vegas area are described in the road logs that follow. Folds result from the flexing or wrinkling of strata, and examples of different types of folds are present in southern Nevada. Upwardly convex folds are called anticlines, and upwardly concave folds are synclines. Folds that are tilted on their sides so that some of the strata are upside down (older rocks overlying younger rocks) are called overturned folds, and are often associated with thrust faults.

Faults, linear breaks in the Earth's crust that form during seismic activity (earthquakes), are common and well exposed in many places in southern Nevada. Faults may be classified into several types on the basis of steepness and the motion along them. Geologists have long recognized low-angle thrust faults and high-angle normal and strike-slip faults as important structural features in the Las Vegas area. Low-angle normal or detachment faults, which commonly have straight or curved normal faults in their upper plates, are significant components of the geologic story, especially in the mountains along the Colorado River southeast of Las Vegas.

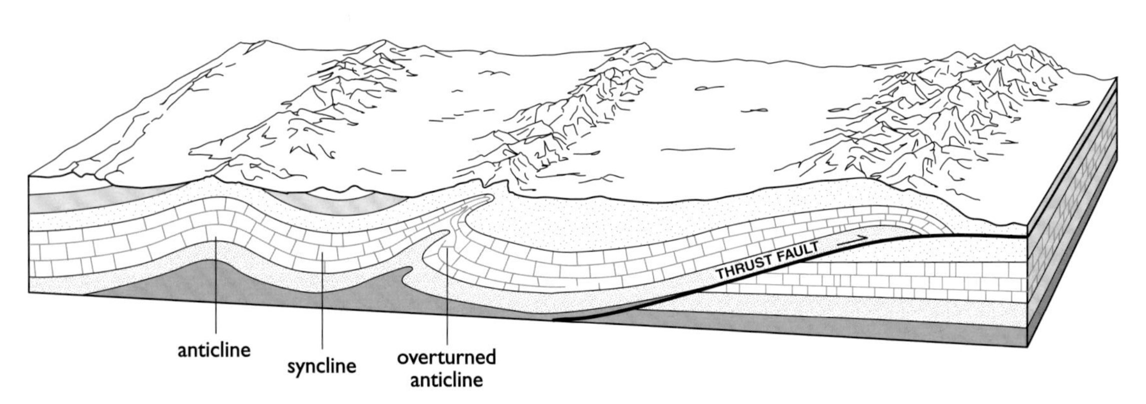

Types of folds showing the progression from gentle folds to overturned folds and eventual thrusting.

Plate Tectonics

Early in the 20th century, on the basis of physical and geologic correlations between continents, some scientists proposed that the Earth's landmasses are mobile, sliding slowly about on the Earth's surface. This overall theory of crustal movement is called plate tectonics.

In plate tectonics theory, new oceanic crust is generated by magma emplaced into the sea floor at spreading ridges. Carpets of this oceanic crust move slowly away from the spreading ridges in a configuration much like conveyor belts. As new crust forms at spreading ridges, it is consumed in areas called subduction zones, most of which are at continental margins. Downward moving (subducted) oceanic crust is pushed into the mantle at continental margins, eventually producing magma that rises through the continental crust to become plutonic or volcanic rocks.

The plates can be very large. For example, much of North America is part of the North American plate. To the west, underlying nearly all of the Pacific Ocean, is the huge Pacific plate.

Until about 20 million years ago, the Farallon plate, an offshore plate lying to the east of the Pacific plate, was moving eastward and colliding directly with the North American plate. As a result, the Farallon plate was forced beneath the North

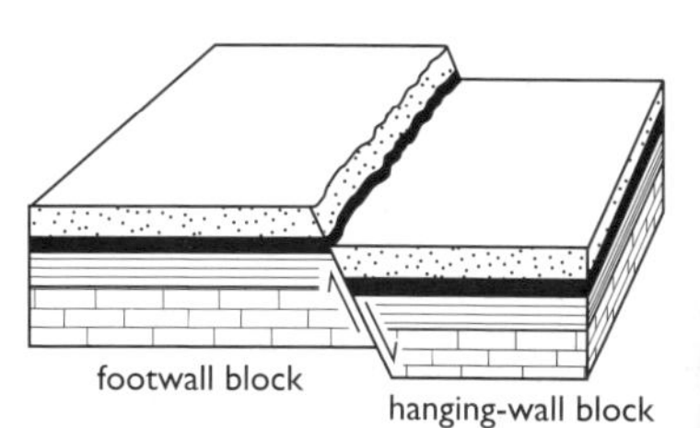

Normal fault

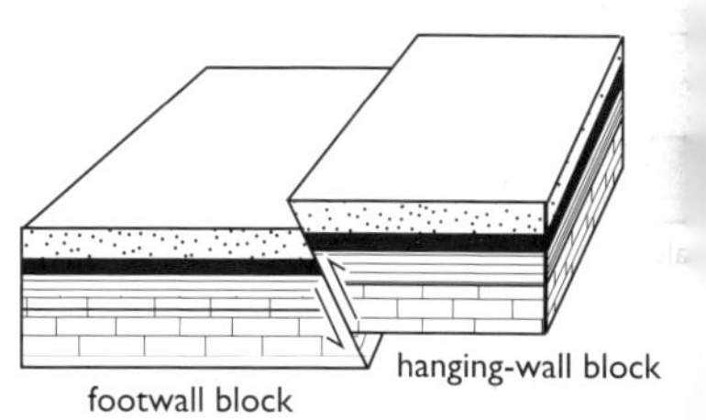

Reverse fault

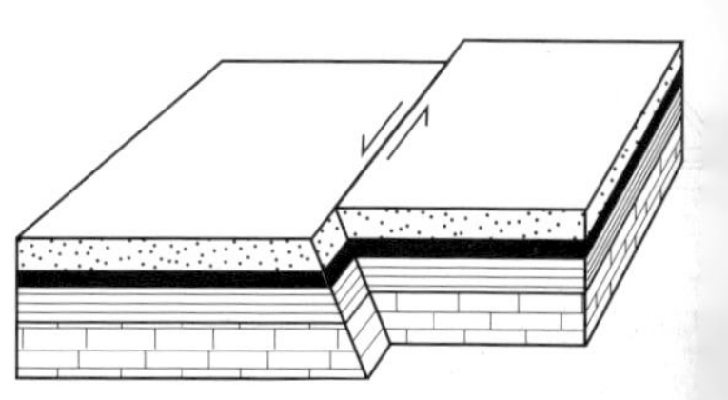
Left-lateral strike-slip fault

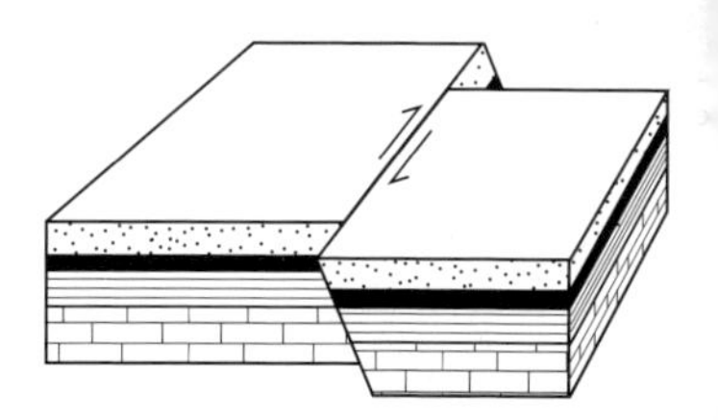
Right-lateral strike-slip fault

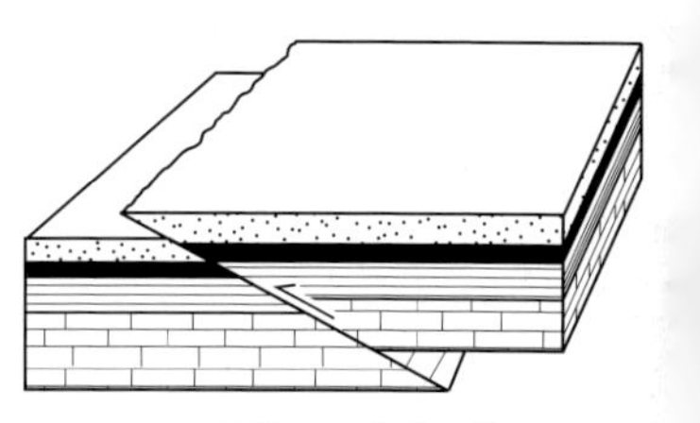
Thrust fault

Types of faults.

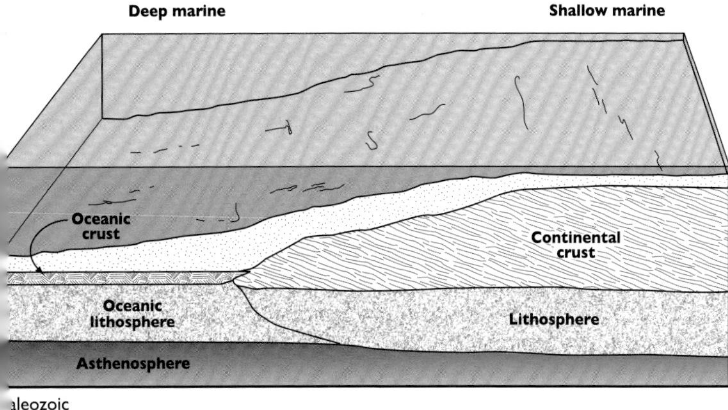

aleozoic

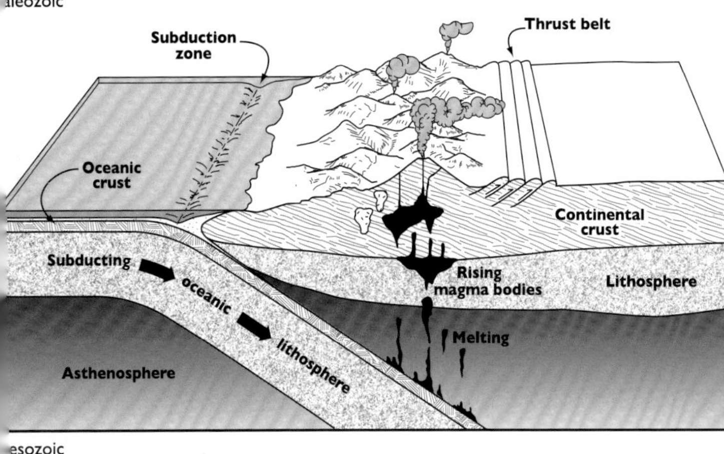

esozoic

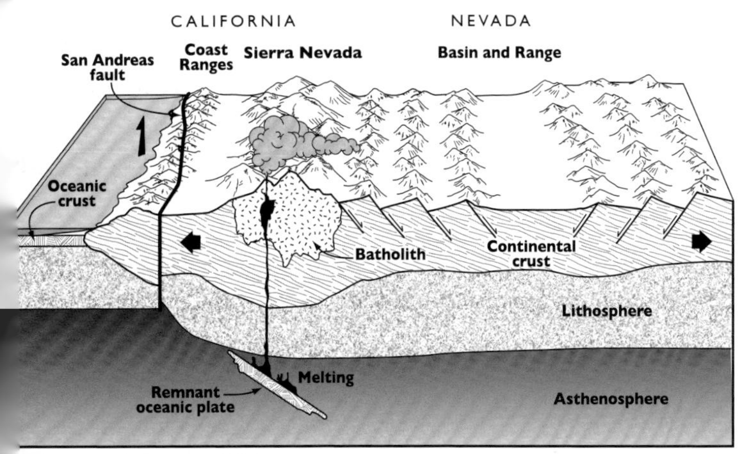

te Cenozoic (from about 20 Ma)

mplified crustal block diagram showing plate tectonic processes at a bduction zone at a continental margin; oceanic crust subducting beneath ntinental crust.

American plate, becoming completely "subducted," leaving the Pacific plate in contact with the North American plate. At the present time, our two remaining local plates are not colliding head-on, but are sliding together laterally along a boundary geologists call the San Andreas fault; the Pacific plate is moving to the northwest in relation to North America.

The present northwestward sliding movement of the Pacific plate tends to create tensional (pull-apart) forces in the North American block. Plate interactions have resulted in northwest-trending, right-lateral faults (parallel to the San Andreas fault) and generally north-trending normal faults that define "fault-block" mountain ranges with down-dropped valleys between them—our Basin and Range province.

Geologic History

Rocks from all of the four major divisions of geologic time, Precambrian (from about 4.5 billion to 540 million years ago), Paleozoic (from 540 million to 248 million years ago), Mesozoic (from 248 to 65 million years ago), and Cenozoic (from 65 million years ago to the present—we live in the Cenozoic) are represented in the Las Vegas area.

To geologists, rock ages are relative. That is, the age of one rock formation is described in relation to the age of the formations above and below it. Sedimentary rocks, undisturbed by folding or faulting, are pretty simple—the rocks on the bottom of the pile were deposited first so they are the oldest. The youngest rocks are the ones on top. Volcanic flows work the same way, oldest on the bottom and youngest on top. When we become familiar with a normal pattern for a stack of rocks, we can recognize them even if they have been messed up by folding and faulting—we know what should be on the bottom—and if we find the bottom rocks on top, we know they have been overturned. Even better, if we employ one of the radiometric dating techniques (where age is determined by measuring natural radioactive decay) we can assign a fairly precise age to rocks.

Precambrian Rocks—The Basement

Geologists generally apply the term basement to Precambrian (older than 540 million years) metamorphic and igneous rocks that make up a complexly deformed footing upon which more ordered sedimentary strata are deposited. The oldest rocks in the Las Vegas area are gneiss and schist that geologists believe were originally deposited as sediments and volcanic rocks 1.7 to 1.8 billion years ago during Precambrian time. About 1.4 billion years ago, large amounts of granitic magma (molten rock) were either forced or melted their way upward into these metamorphic rocks during a period of possible continental rifting (breaking apart and spreading) and widespread igneous activity in the ancestral North American continent.

Exposures of Precambrian metamorphic and igneous rocks occur to the east and south of Las Vegas and these rocks can be seen on Trips 3 and 4.

The Paleozoic—Tropical Seas

During the Paleozoic Era (from 540 to 248 million years ago), southern Nevada and much of the eastern Great Basin region were at the edge of the North American continent. The earliest Paleozoic rocks deposited on this continental shelf setting were clastic sedimentary rocks. In what is now the Las Vegas area, Paleozoic seas lapping on the Precambrian bedrock of the continent deposited beach deposits of the 520-million-year-old Tapeats Sandstone. This was followed by the deposition of the Pioche Shale, which locally contains abundant early Cambrian trilobite fossils. However, the Paleozoic in southern Nevada was mainly an era of geologic quiescence and nonclastic sedimentary deposition, marked by slow accumulation of limestone and dolomite in ancient shallow tropical seas that were similar to shallow parts of the present-day Caribbean. Beds of particularly pure Paleozoic limestone and dolomite are mined and processed into lime near Las Vegas.

Late in the Paleozoic, during the Permian Period, limestone deposition gave way to deposition of sandstone and gypsum. The Toroweap and Kaibab Formations, which mainly consist of limestone with abundant brachiopod and crinoid fossils, also contain sandstone and shale with associated gypsum beds. Gypsum in the Kaibab Formation is mined near Las Vegas and used in manufacturing wallboard.

Outcrops of Paleozoic rock can be observed on Trips 1, 2, and 3.

The Mesozoic—Forests and Deserts

The Mesozoic Era (from 248 to 65 million years ago), the age of the dinosaurs, was marked in southern Nevada by a change from marine to continental sedimentary deposition. Rocks in the lower part of the Triassic Moenkopi Formation contain fossils that indicate marine deposition, and these mark the last time that southern Nevada lay under the ocean. The Moenkopi was succeeded by the Chinle Formation, mainly clastic continental rocks that contain remains of ancient forests. Chinle deposition was followed by deposition of the Aztec Sandstone, a thick accumulation of wind-blown desert sand, whose fiery red outcrops occur to the west and east of Las Vegas and can be seen on Trips 1 and 3. The youngest Mesozoic formation, the Cretaceous Baseline Sandstone, is mined to produce silica sand near Overton.

The Mesozoic was also a time of significant regional change in the configuration of the Earth's crust. Beginning in the early Mesozoic, easterly moving rocks of the Pacific oceanic crust began to descend beneath the westerly moving North American continent along a subduction zone. This resulted in intense magmatic activity that produced the granitic rocks of the Sierra Nevada and also caused compression, or horizontal squeezing, of rocks to the east that were previously laid down in the Paleozoic sea. The Mesozoic in southern Nevada was an era of uplift, large-scale folding of sedimentary rock strata, and thrust faulting caused by this compression. Thrust faults are seen on Trips 1, 2, and 3.

The Cenozoic—A Tug-of-War

During the Cenozoic Era (from 65 million years ago to the present), the Las Vegas area underwent a second major change in the style of geologic movement. The Pacific oceanic crust changed its movement from collision and subduction to right-lateral shearing. This oblique movement signaled the beginning of extensional deformation in the area west of the Colorado Plateau. This area began to be literally pulled apart. The rocks were tilted, folded, and broken by faults during this deformation. Geologists believe that rocks that make up the Sierra Nevada were much closer to those of the Colorado Plateau in the early part of Cenozoic time than they are today. In southern Nevada, total east-west extensional movement is estimated at about 150 miles. During the latter part of the Cenozoic (about the last 20 million years), extensional forces produced the mountain ranges and intervening valleys that are characteristic of the Basin and Range province. Nearly all mountain ranges in Nevada are bounded by at least one fault, and these faults are considered to be active, moving during major earthquakes every few thousand years.

Crustal extension in southern Nevada resulted in the deposition of clastic and nonclastic sedimentary rocks of the Horse Spring Formation in fault-bounded basins in the Lake Mead area between about 17 and 10 million years ago. Horse Spring Formation strata were later tilted by extensional faulting, but clastic sedimentary rocks of the 10- to 5-million-year-old Muddy Creek Formation (which overlies the Horse Spring Formation) are largely undeformed. This indicates that extensional faulting and related deformation had significantly slowed down, or even come to and end by the time that the Muddy Creek Formation was deposited. Volcanic activity also occurred between about 17 and 5 million years ago in southern Nevada as magma rose from deeper in the crust and upper mantle along vents and fissures in the extensionally thinned crust. Some of the effects of crustal extension are seen on Trip 3, and volcanic rocks and landforms are evident on Trip 4.

Major Divisions of Geologic Time

GEOLOGIC AGE				Lower boundary		DOMINANT LIFE
CENOZOIC	Quaternary	Neogene	Holocene	0.01 Ma	The several geologic eras were originally named Primary, Secondary, Tertiary and Quaternary. The first two names are no longer used. Tertiary and Quaternary have been retained and used as period designations.	
			Pleistocene	1.8 Ma		
	Tertiary		Pliocene	5.3 Ma		
			Miocene	23.8 Ma		
		Paleogene	Oligocene	33.7 Ma		
			Eocene	54.8 Ma		
			Paleocene	65 Ma		
MESOZOIC	Cretaceous			144 Ma	Derived from Latin word for chalk (creta) and first applied to extensive deposits that form white cliffs along the English Channel.	
	Jurassic			206 Ma	Named for the Jura Mountains, located between France and Switzerland, where rocks of this age were first studied.	
	Triassic			248 Ma	Taken from word "trias" in recognition of the threefold character of these rocks in Europe.	
PALEOZOIC	Permian			290 Ma	Named after the ancient Kingdom of Permia in Russia, where these rocks were first studied.	
	Pennsylvanian			323 Ma	Named after the state of Pennsylvania, where these rocks have produced much coal.	
	Mississippian			354 Ma	Named for the Mississippi River valley, where these rocks are well exposed.	
	Devonian			417 Ma	Named after Devonshire, England, where these rocks were first studied.	
	Silurian			443 Ma	Named after Celtic tribes, the Silures and Ordovices, that lived in Wales during the Roman Conquest.	
	Ordovician			490 Ma		
	Cambrian			540 Ma	Taken from Roman name for Wales (Cambria) where rocks containing the earliest evidence of complex forms of life were first studied.	
PRECAMBRIAN	Proterozoic			2,500 Ma	The time between the birth of the planet and the appearance of complex forms of life. More than 80 percent of the Earth's estimated 4.5 billion years is Precambrian.	
	Archean			4,550 Ma		

Ma = Mega-annum = million years

(data from Geological Society of Americ

Mineral Deposits

Exploitation of minerals has been an important part of human activity in southern Nevada dating back to the Ancient Puebloan salt miners over 1,000 years ago. Spanish adventurers may have mined gold near Nelson in Eldorado Canyon in the 18th century, and Mormon settlers mined lead and silver from deposits in Lucky Strike Canyon in the Spring Mountains in 1856, predating mining on northern Nevada's Comstock Lode by three years.

The largest metal mining operations in this part of southern Nevada were for gold at Eldorado Canyon in the late 19th and early 20th centuries and for manganese at the Three Kids Mine east of Las Vegas between 1917 and 1961. There are no metal mines currently in operation in the area explored by our trips. Sites of some of the former operations can be seen on Trip 4, and we pass close to some old mine sites on Trips 2 and 3.

Today, nonmetallic minerals (commonly referred to as industrial minerals) are the most important mineral commodities sought in southern Nevada. The borate and magnesite mines are not active, and the salt deposits are largely inundated by Lake Mead, but large deposits of gypsum, silica, limestone, and construction aggregate are being mined from several locations in Clark County. You will see large gypsum mines on Trips 1 and 3, you will pass close by silica and limestone mines on Trip 3, and you will see a large construction aggregate quarry at the end of Trip 4. Sand and gravel, essential for use in the booming construction industry in the Las Vegas area, are mined from pits in several locations in Las Vegas Valley. Lone Mountain, one of these locations, lies west of U.S. Highway 95 and can be seen as you travel north from Las Vegas toward the starting point of Trip 2.

Please keep in mind that mines, both old and new, are best viewed from a distance. Old mines are very dangerous and should not be explored under any conditions. Active mining operations are also dangerous and are no place for tourists to wander about unescorted. Some mine operators, however, conduct special tours for schools and other groups.

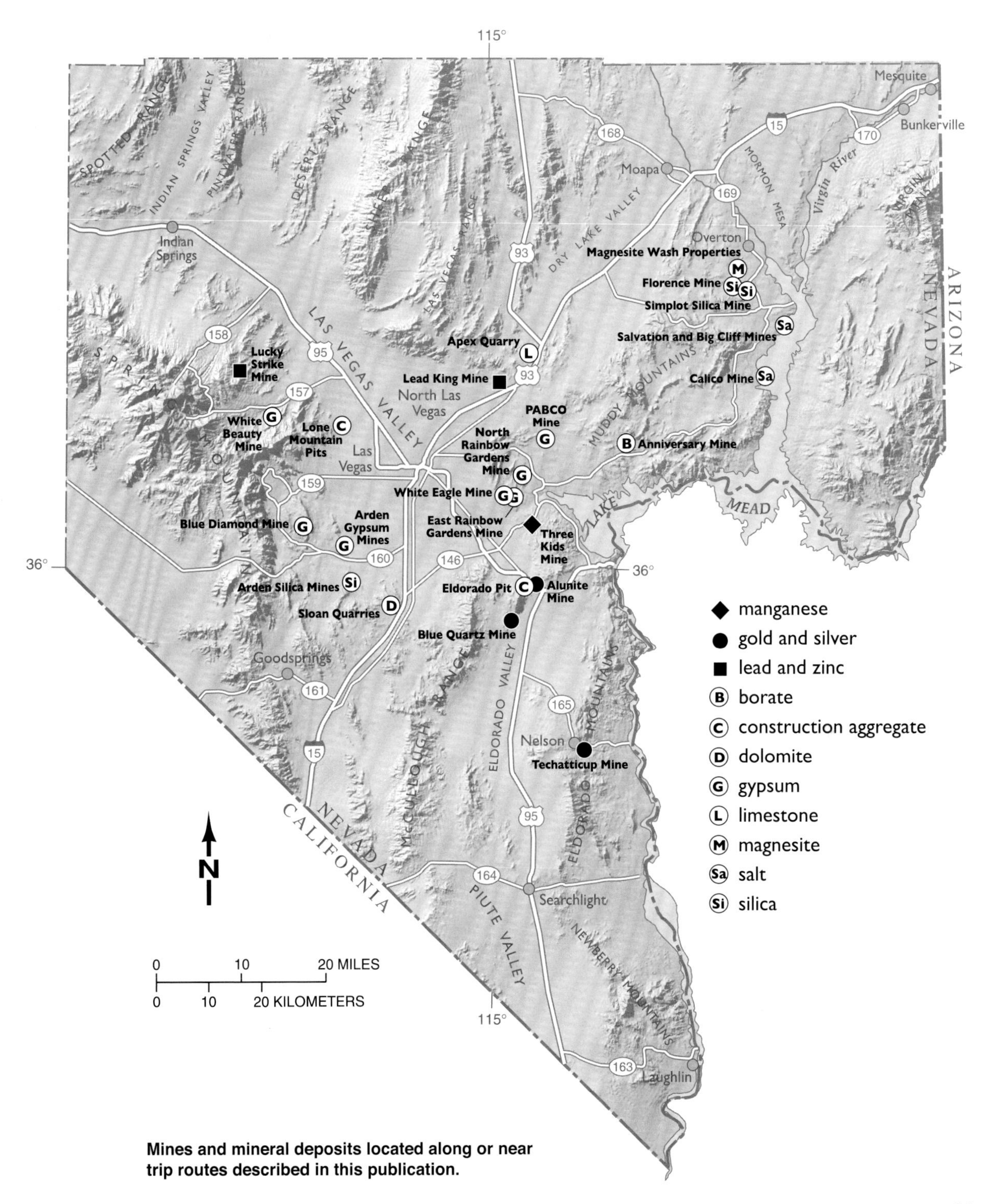

Mines and mineral deposits located along or near trip routes described in this publication.

COLOR PHOTO CAPTIONS

PLATE 1

Generalized geology of central Clark County, Nevada.

PLATE 2

2a **Desert bighorn sheep** (*Ovis canadensis*) Rams 5 to 6 feet long, ewes 4 to 5 feet long; sturdy, muscular build; brown to tan with white belly, rump, muzzle, and eye patches. (Rams shown)

Rams have massive horns that grow over and behind the ears in a C-shaped "curl." Horns are permanent and grow incrementally year by year. It may take 7 to 8 years for a ram to acquire a full curl. Ewes have shorter, more slender horns that curve gently back.

Bighorn sheep occupy rough, precipitous terrain near sources of permanent water. Bands follow regular feeding routes and maintain bedding grounds that may be used for years. The desert bighorn sheep is Nevada's state mammal. *Photo: Red Rock Canyon National Conservation Area*

2b View to the south across **Red Rock Wash** from the summit of Turtlehead Mountain, Red Rock Canyon National Conservation Area. The Sandstone Bluffs are in the background. *Photo: Mark Vollmer*

2c Now called "wild," the **burros** that you might see roaming about the Red Rock Canyon area are descendents of hard-working pack animals introduced into the southwest by Spaniards in the 16th century. Used by early miners and other travelers who passed through this desert country, the burros were set free by their owners when they either had no further use for them, or just lost patience with the animal's somewhat independent personality. Because they have no natural enemies and are very hardy, the burros have thrived throughout the North American deserts and compete for limited food and water resources with native animals such as desert bighorn sheep. *Photo: Becky Purkey*

2d Springtime bloom of **Phacelia** (*Phacelia spp.*) in Red Rock Canyon National Conservation Area. *Photo: Joe Tingley*

2e **Prickly poppy** (*Argemone platyceras*) grows in dry, sandy and gravelly soils and is commonly seen along roadsides. It grows to 3 feet tall with stems and leaves covered with sharp spines. The plant contains alkaloids toxic to man and other animals. *Photo: Carol McKim*

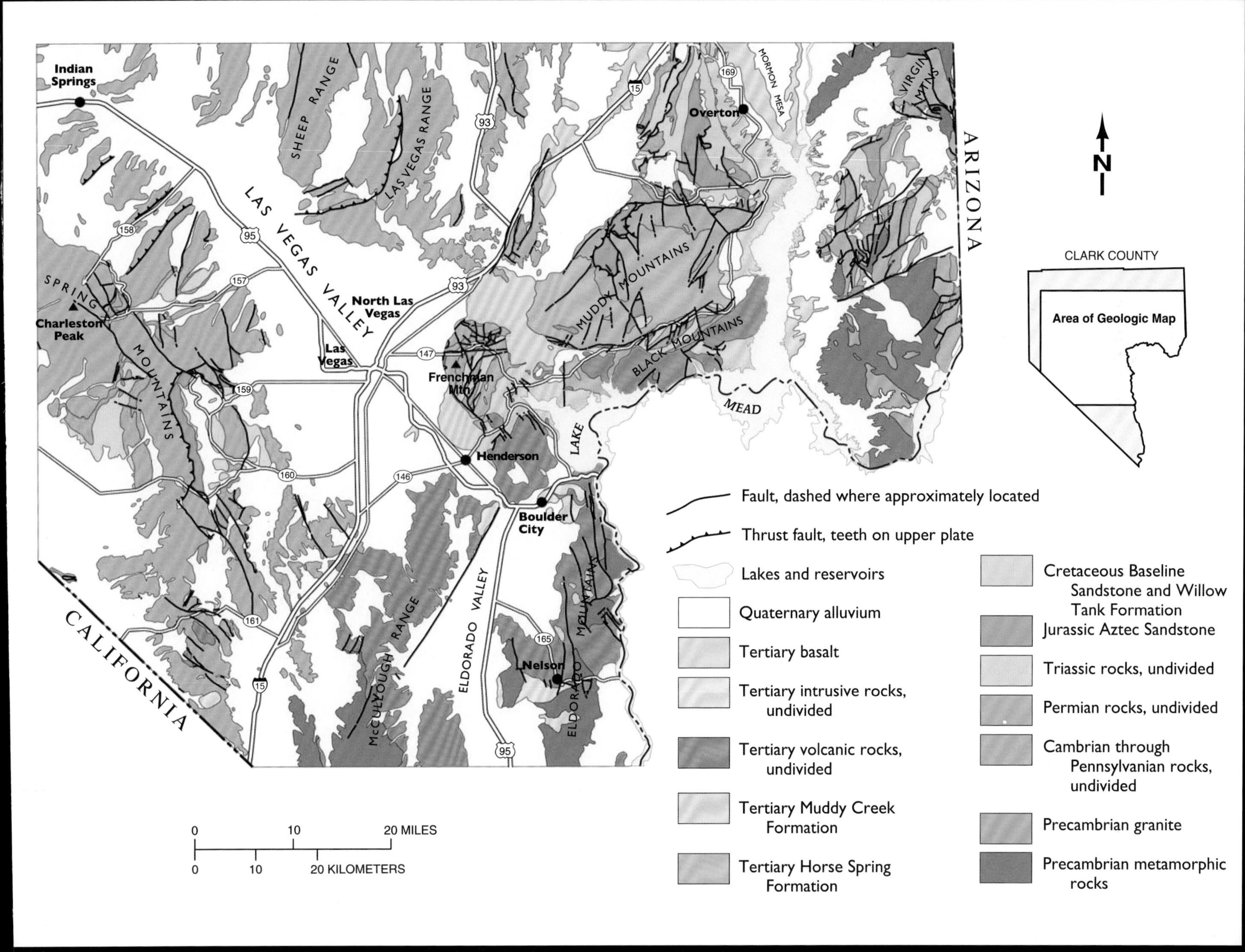
Indian Springs
SHEEP RANGE
LAS VEGAS RANGE
MORMON MESA
Overton
VIRGIN MTNS
ARIZONA
N
LAS VEGAS VALLEY
SPRING MOUNTAINS
Charleston Peak
North Las Vegas
Las Vegas
Frenchman Mtn
MUDDY MOUNTAINS
BLACK MOUNTAINS
LAKE MEAD
Henderson
Boulder City
CLARK COUNTY
Area of Geologic Map
McCULLOUGH RANGE
ELDORADO VALLEY
ELDORADO MOUNTAINS
Nelson
CALIFORNIA
15
93
95
147
157
158
159
160
146
161
165
169
0
10
20 MILES
0
10
20 KILOMETERS
Fault, dashed where approximately located
Thrust fault, teeth on upper plate
Lakes and reservoirs
Quaternary alluvium
Tertiary basalt
Tertiary intrusive rocks, undivided
Tertiary volcanic rocks, undivided
Tertiary Muddy Creek Formation
Tertiary Horse Spring Formation
Cretaceous Baseline Sandstone and Willow Tank Formation
Jurassic Aztec Sandstone
Triassic rocks, undivided
Permian rocks, undivided
Cambrian through Pennsylvanian rocks, undivided
Precambrian granite
Precambrian metamorphic rocks

PLATE 1

2a

2b

2c

2d

2e

PLATE 2

PLATE 3

PLATE 4

Color Photo Captions

PLATE 3

3a View of the **White Rock Hills** from the White Rock Spring Trail, Red Rock Canyon National Conservation Area. The Aztec Sandstone in the picture is in the footwall of the Keystone thrust, and the rocks were folded into a syncline by the force of the thrusting. Notice the beds on the left (the white rock) dip gently to the left while the beds on the right (mostly all red rock) are nearly vertical. *Photo: Joe Tingley*

3b Deceptively fragile-looking **desert gold poppy** (*Eschscholtia glyptosperma*). *Photo: Patti Gray*

3c The beginning of fall color in the cottonwoods at **Willow Spring Picnic Area**, Red Rock Canyon National Conservation Area. *Photo: Becky Purkey*

3d A view of **Calico Basin** from Red Spring, Red Rock Canyon National Conservation Area. The gray rock on the eastern skyline in the background is Cambrian Bonanza King limestone and dolomite in the upper plate of the Keystone thrust. Gray talus cones formed from the eroding Cambrian rock cascade down across the red Aztec Sandstone in the lower plate of the thrust. *Photo: Mark Vollmer*

3e Red and white cross-bedded **Aztec Sandstone** in the Calico Hills, Red Rock Canyon National Conservation Area. *Photo: Mark Vollmer*

3f **Ancient Puebloan petroglyphs** keep watch over Red Spring in the Red Rock Canyon National Conservation Area. The spring area, a historical Ancient Puebloan/Paiute camp, is now a picnic area and a habitat restoration project within the Red Rock Canyon National Conservation Area. Private development in Calico Basin outside of the Red Rock Canyon National Conservation Area is in the distance. *Photo: Mark Vollmer*

PLATE 4

4a A **Joshua tree** (*Yucca brevifolia*) frames the snow-covered Spring Mountains; inset is of another Joshua tree in bloom. The Joshua tree is the largest of the yuccas and is the characteristic tree of the Mojave desert. *Photo: Mark Vollmer; inset: Joe Tingley*

4b **Prince's plume** (*Stanleya pinnata*). A member of the mustard family, prince's plume is found in sandy soils and talus slopes extending from the deserts to the lower mountains. *Photo: Joe Tingley*

4c White **yucca** flowers stand out in contrast to the blue desert sky. *Photo: Jack Hursh*

4d **Paintbrush** (*Castilleja spp.*) Because of hybridization and wide intraspecific variation, many species of paintbrush are hard to tell apart. Flower colors range from brilliant red to orange to yellow. Paintbrush can be found from the low bajadas to above the timberline. *Photo: Jack Hursh*

4e **Painted lady butterfly** (*Cynthia cardui*) on **brittlebush**. *Photo: Jack Hursh*

4f **Prickly pear cactus** (*Opuntia polycantha*) forms low, spiny mounds seldom more than 6 inches high. It prefers sandy soils and can be found in desert areas as well as piñon-juniper woodlands. Prickly pear flowers may also be magenta. *Photo: Jack Hursh*

interval | cumulative

TRIP 1: BLUE DIAMOND AND RED ROCK CANYON NATIONAL CONSERVATION AREA

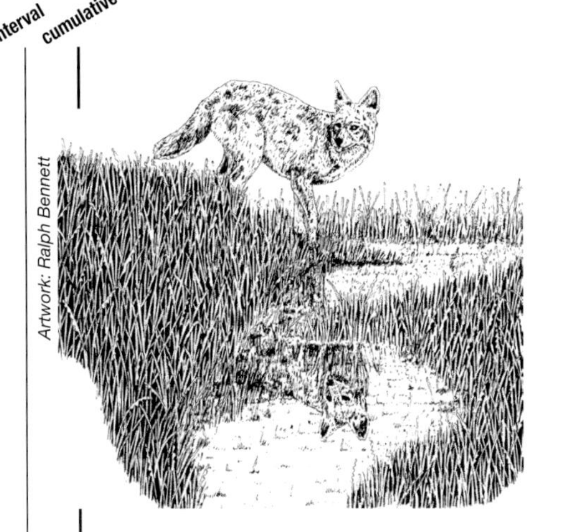
Artwork: Ralph Bennett

This trip takes you west of Las Vegas to the southern Spring Mountains where you will travel first to the small town of Blue Diamond, then past the spectacular Sandstone Bluffs and into the Bureau of Land Management's Red Rock Canyon National Conservation Area. The Sandstone Bluffs are also known as the Red Rock Canyon Sandstone Escarpment or the Wilson Cliffs. The escarpment, formed on colorful red and white rocks of the Jurassic Aztec Sandstone, defines about 12 miles of the eastern face of the Spring Mountains. Along and leading into the bluffs there are hiking trails to springs and hidden canyons, and there are ancient petroglyphs and agave roasting pits, wild burros, and Joshua tree forests to study, see and explore. Geologic highlights include close-up observations of the Aztec Sandstone and the thrust faults that occurred in this area on a gigantic scale during late Mesozoic time, and views of an operating gypsum mine. The trip is about 50 miles in length. Gasoline is available at several locations near the start of the route. Water is available at several places along the way including Blue Diamond, Spring Mountain Ranch State Park, and the visitor center at Red Rock Canyon National Conservation Area.

0.0 — From downtown Las Vegas, drive south on Interstate 15, take Exit 33 West and turn right onto State Route 160. Set odometer at 0.0 at the intersection of S.R. 160 (Blue Diamond Highway) and Dean Martin Drive and continue west on S.R. 160. *(GPS 1, see appendix, p. 141)*

2.5 | 2.5 — Railroad crossing. Arden, a former station on the Union Pacific railroad, is about one mile south (to the left) along the tracks. In 1929, the first known petroleum exploration well in Clark County was drilled a few miles to the northwest of here in a geologic structure called the Arden dome. This dome is an anticline whose axis is bowed-up—all of the rock layers dip away from the high point. This hole, the E.W. Bannister No. 1 well, only went to 552 feet. A later test, the Commonwealth Oil Company No. 1 well drilled in 1933, went to a depth of 1,897 feet. This well, and another even deeper well drilled in 1944, reported traces of oil but nothing of commercial interest was found. *(GPS 2)*

1.7 | 4.2 — The small hill on the left is Exploration Peak. The name is new, made official by the U.S. Board on Geographic Names in 2005, to commemorate early explorers who traveled the nearby Old Spanish Trail between Santa Fe and Los Angeles. The resistant gray rock layers seen on the hill are limestone beds of the Permian Kaibab Formation which dip gently to the northeast. During the "Arden Dome" oil excitement, promoters are said to have led potential investors onto this hill to point out rock dipping to the east, then south, then west as the rock layers wrapped themselves around the hill, indicative of a dome—a good "oil play." *(GPS 3)*

0.4 | 4.6 — The Sandstone Bluffs are straight ahead. Mount Wilson (7,070 feet) is the high peak to the right at 1:00 and the prominent ridge to the left of the highway at 11:00 is Potosi Mountain (8,512 feet). It was on Potosi Mountain that a TWA flight from Las Vegas to Los Angeles crashed on the evening of January 17, 1942, killing all on board including film star Carol Lombard. The actress, who was married to Clark Gable, was returning to California from a defense bond campaign in her home state of Indiana.

GEOLOGIC PERIOD	FORMATION NAME	CORRELATING UNIT ON GEOLOGIC MAP (see page 13)
Quaternary	**Alluvial deposits**	Quaternary alluvium
1.8 Ma		
Tertiary	**Landslide breccia**	not shown
65 Ma	***unconformity***	
Jurassic	**Aztec Sandstone** (red, tan, or yellow cross-bedded sandstone)	Jurassic Aztec Sandstone
206 Ma		
Triassic	**Kayenta Formation** (brick red sandstone and siltstone) **Chinle Formation** (dark red to purple siltstone, mudstone, minor sandstone, and conglomerate) **Moenkopi Formation** (reddish siltstone, shale, and sandstone, with a prominent limestone member - the Virgin Limestone)	Triassic rocks, undivided
248 Ma	***unconformity***	
Permian	**Kaibab and Toroweap Formations, undivided** (gray limestone with intervals of white gypsum) **Permian red beds** (red sandstone and siltstone)	Permian rocks, undivided
290 Ma		
Pennsylvanian	**Bird Spring Formation** (gray limestone and dolomite)	Cambrian through Pennsylvanian rocks, undivided
323 Ma		
Mississippian	**Monte Cristo Limestone** (gray limestone)	
354 Ma		
Devonian	**Sultan Limestone** (gray limestone)	
	unconformity, Silurian rocks missing	
Ordovician	**Pogonip Group, Eureka Quartzite, and Ely Springs Dolomite, undivided** (dolomite, calcareous shale, quartzite)	
490 Ma		
Cambrian	**Nopah Formation and Dunderberg Shale** (dark gray shale and interbedded dolomite) **Bonanza King Formation** (gray dolomite and limestone)	
540 Ma		

Numbers followed by "Ma" indicate millions of years ago.

Generalized stratigraphic column for the Red Rock area and the Spring Mountains

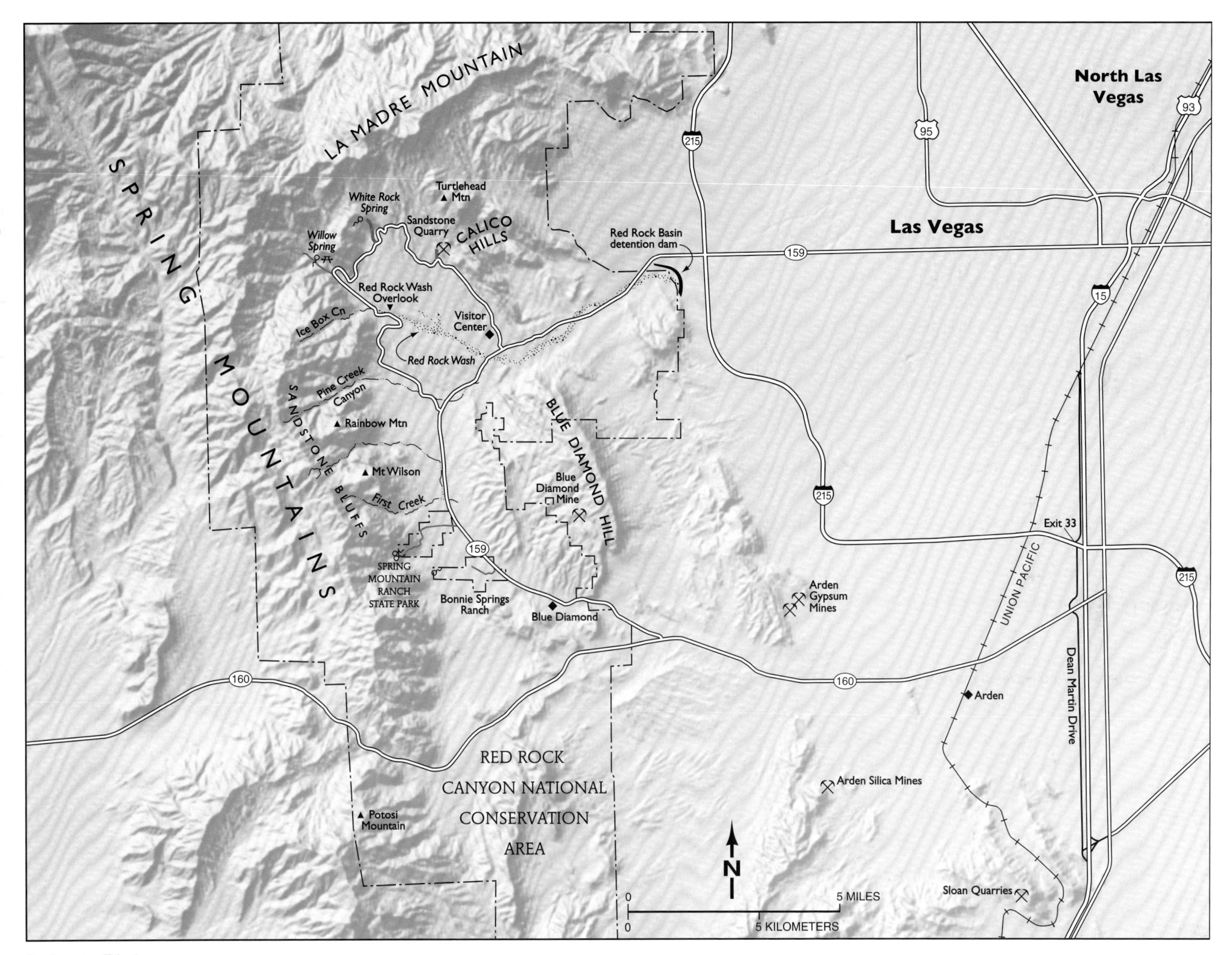

Route map, Trip 1.

THE KEYSTONE THRUST SYSTEM

The Sandstone Bluffs, the 2,000-foot-high cliffs in view throughout most of Trip 1, are composed dominantly of red and white Aztec Sandstone (upper right photo on the front cover). The Aztec Sandstone is Jurassic in age (probably deposited about 180 million years ago) and is the same rock formation as the Navajo Sandstone found on the Colorado Plateau.

On the skyline you can see a series of dark gray rock layers that now lie above the Aztec Sandstone. These strata are limestone and dolomite of the Cambrian Bonanza King Formation. The sequence of rocks exposed in this escarpment is out of normal stratigraphic order; that is, older rocks are resting upon younger rocks (Cambrian rocks on Jurassic rocks, see geologic time chart on page 10). Since this unusual arrangement cannot result from normal depositional processes, it indicates the presence of a major fault—in this case a thrust fault—between the two layers.

Faults that generally place older rocks on top of younger rocks are known as reverse faults and, if they are inclined at low angles, they are called thrust faults. Thrust faults result from compressional forces within the Earth's crust.

This particular thrust fault is known as the Wilson Cliffs thrust, one of the oldest (here topographically lowest) of several west-dipping, low-angle faults which make up the well-known Keystone thrust system. The fault is inclined gently downward to the west and places the older Cambrian rocks on top of younger Jurassic rocks. Because of its spectacular exposure—in many places it appears as a knife-sharp contact from the air—this structure is one of the most photographed thrust faults in the world.

The Keystone thrust system is one of a series of thrust faults in southern Nevada that collectively moved blocks of rock—originally 3 to 4 miles thick—tens of miles eastward. Movement along these great thrust faults probably occurred during the middle of the Cretaceous Period, although the exact age of activity on these faults remains a point of controversy among geologists.

younger rocks

older rocks

unfaulted

high-angle reverse fault

low-angle reverse (or thrust) fault

Aerial view to the northwest along the Sandstone Bluffs. The trace of the Wilson Cliffs thrust is marked by the abrupt color change from light to dark at the top of the bluffs. The Keystone thrust, parallel to the Wilson Cliffs thrust but not marked by a distinct color change, is within the dark rocks to the left. ▶

Photo: Becky Purkey

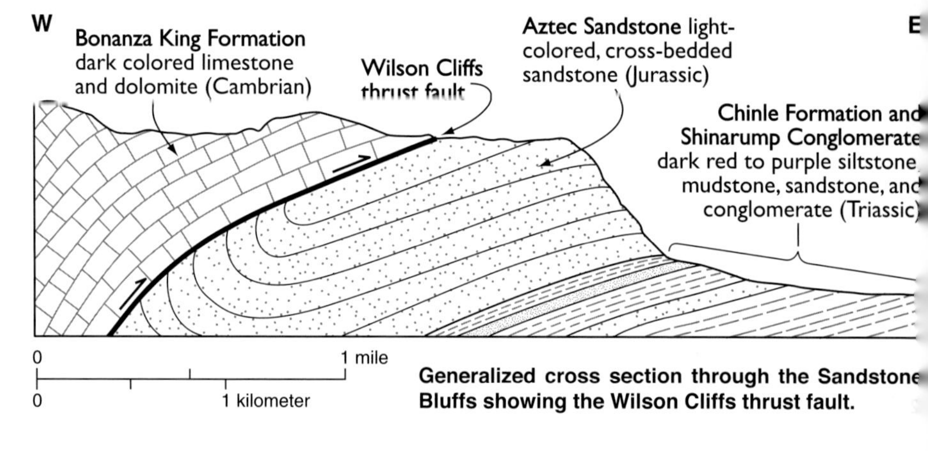

Generalized cross section through the Sandstone Bluffs showing the Wilson Cliffs thrust fault.

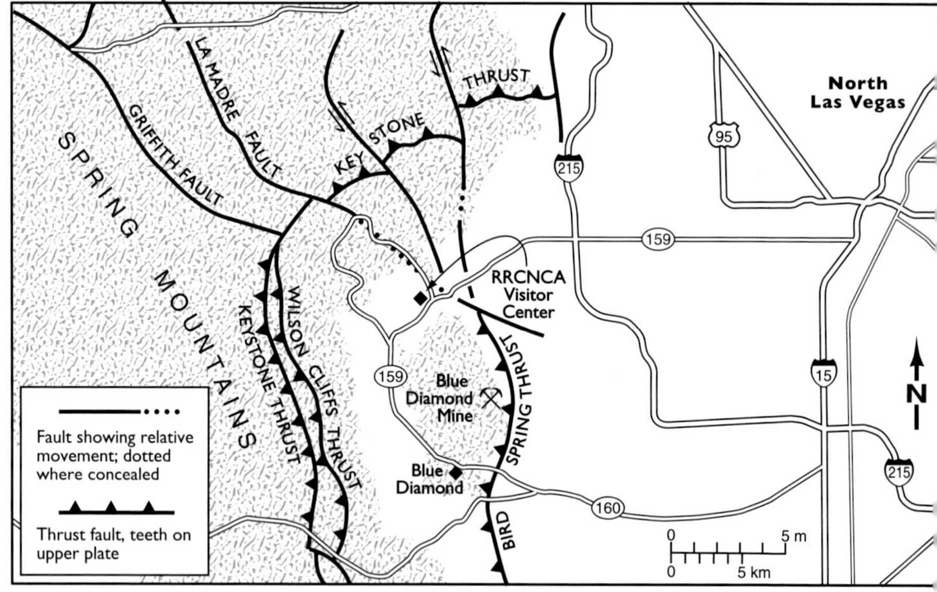

Major faults in the southern Spring Mountains (after Page and others). ▶

interval	cumulative
1.2	5.8

Although development has erased any surface trace of it, the site of J.B Nelson No. 1 Oil Well, drilled on the west flank of the Arden Dome was about 100 feet south of the highway at this point. Drilling began in 1943, but the well was abandoned at 2,210 feet when the drilling tools twisted off in the hole.

At the foot of the low range of hills about 2½ miles farther to the south, are the Arden (also known as Bard) silica mines. Silica was mined from a Permian sandstone unit and used for casting sand (sand used to make molds) in foundries.

0.2 | 6.0

At about 2:00, note the conical hill with reddish rock skirting it. Also note the cave-like openings slightly above the base of this hill. These are openings into the old, underground Arden gypsum mines. The gypsum occurs in a specific layer in the Kaibab Formation, and you can see that the mine openings all seem to line up along the face of the hill. Now look to the west and you will see a similar line of mine workings along the face of the larger hill. These mines are dug into the same rock formation on the western hill. There are some faults in the valley between the two hills, and the rock layers that should have connected across the gap between the hills have been dropped down and are now covered by gravel in the wash (view from "6.0" on figure, page 22). Even on this small scale, this is a good example of basin-and-range faulting, where fault-bounded blocks have been down-dropped (grabens) between adjacent stationary or possibly uplifted blocks (horsts). *(GPS 4)*

The Arden gypsum mines were active from about 1909 to 1931. When they were in operation, a narrow gauge railroad 5 miles long connected the mines to a plaster mill at Arden on the Union Pacific Railroad.

2.1 | 8.1

Note prominent west-dipping bedding in the limestone ridge top on the right at about 3:30. This ridge is capped with limestone of the Kaibab Formation. Now look ahead (still on the right) at the low, striped hills about 1:30. These striped rocks are the Virgin Limestone Member of the Triassic Moenkopi Formation. There is a steep normal fault in the wash between the two sets of low hills and the rocks to the west have been dropped down in relation to the rocks to the east (view from "8.1" on figure, page 22). *(GPS 7)*

Blue Diamond Hill is beyond the striped hills, extending from about 1:00 to 1:30.

Is It Sandstone Bluffs or Wilson Cliffs?

The escarpment we have been observing and describing is officially named Sandstone Bluffs. By official, we mean Sandstone Bluffs is the name recognized by the U.S. Board on Geographic Names, the naming branch of the U.S. Geological Survey. Government maps, being official and all that, only use "official names" so any published government map will show this feature labeled as "Sandstone Bluffs." But what about the other names, Wilson Cliffs and Red Rock Escarpment? These are alternate names for the bluffs and each has a loyal following of local users. Who named it Sandstone Bluffs? No one knows, but about all this name tells us is that the bluffs are made of sandstone. "Red Rock Escarpment" sounds impressive, and the feature is definitely impressive, but it took little thought to come up with this name for an escarpment cut in red rocks. Wilson Cliffs, on the other hand, isn't descriptive, but this name does have a tie to local history. James Wilson and his partner, George Anderson, established a ranch at the base of the cliffs in 1876 and the Wilson family owned the ranch until 1929. It therefore seems fitting that the rugged cliffs rising to the west of the ranch carry the name Wilson Cliffs. The Wilsons, however, called their holdings "Sandstone Ranch," so could they be the ones who coined "Sandstone Bluffs" as a matching name? The ranch, after several name transformations of its own, is now Spring Mountain Ranch State Park.

▼ **View to the northwest toward the Sandstone Bluffs. Mount Wilson is in the center background of the photo and First Creek Canyon is on the left of Mount Wilson.**

Photo: Joe Tingley

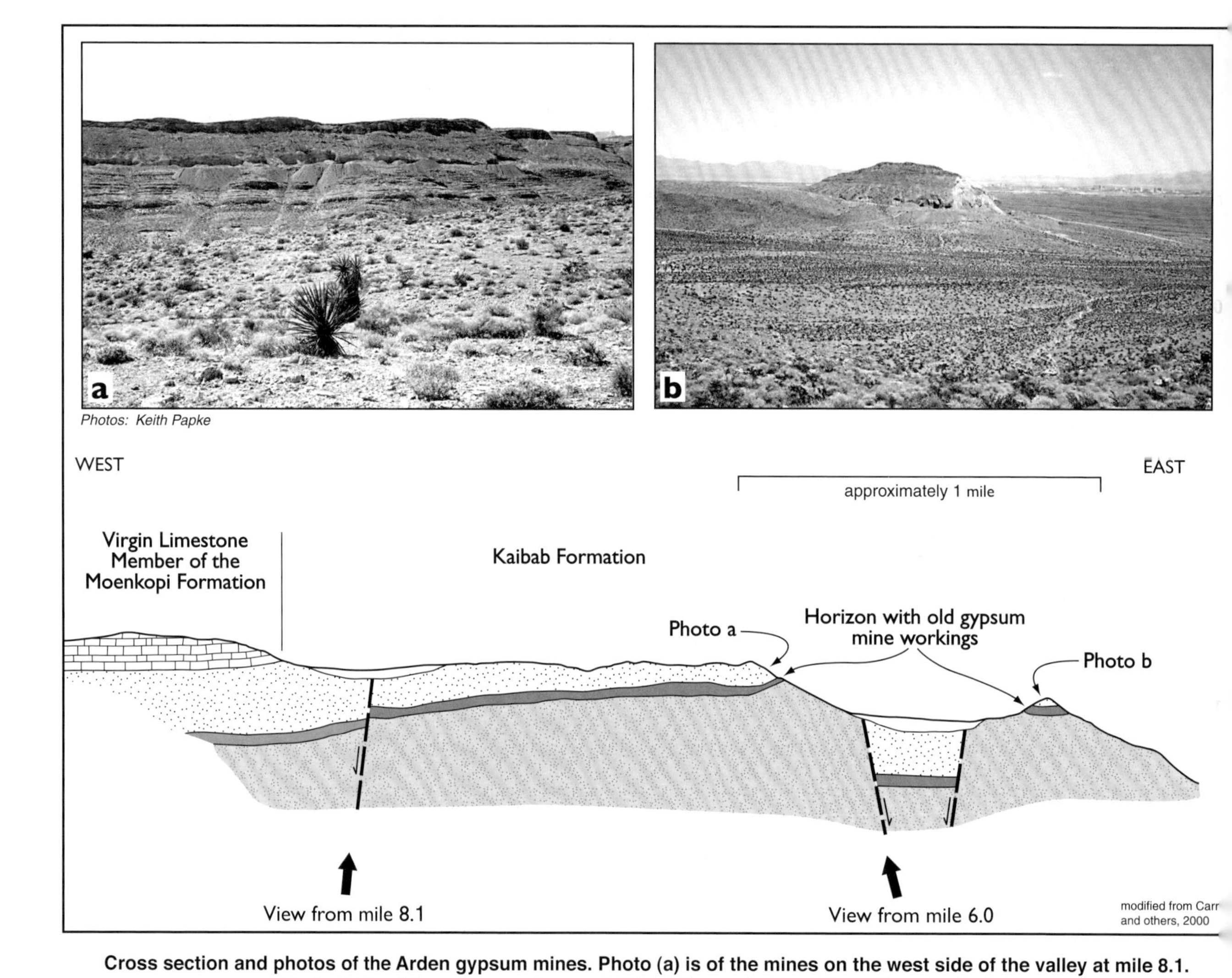

Cross section and photos of the Arden gypsum mines. Photo (a) is of the mines on the west side of the valley at mile 8.1. Photo (b) is of the mines on the small hill on the east side of the valley at mile 6.0.

Steam tractor, courtesy of Red Rock Canyon National Conservation Area.

interval	cumulative	
1.0	9.1	Blue Diamond Hill is at 2:00. The rock capping Blue Diamond Hill is limestone of the Kaibab Formation. The white structure centered on the top of the hill is the top of the ore conveyor system of the Blue Diamond gypsum mine.
0.9	10.0	Intersection with S.R. 159. Turn right (north) toward Blue Diamond and Red Rock Canyon. *(GPS 9)*
0.6	10.6	The Blue Diamond gypsum processing plant and mine are ahead on the right. The route continues west through a gap in strata that are dipping gently (about 10 degrees) to the west. Because of the westward tilt of these strata, one encounters successively younger rocks to the west. For example, the rocks in the vicinity of the gypsum mine are Permian (290 to 248 million years old); the Aztec Sandstone exposed in the escarpment to the west beyond the gap is Jurassic, at least 100 million years younger. Mount Wilson is the prominent peak on the skyline to the right.
1.2	11.8	Turnoff to the Blue Diamond processing plant. Note the recent landslide in the cliff face to the left of the plant site. The more massive limestone unit above has collapsed due to undercutting of a less resistant lower unit. *(GPS 10)*

BLUE DIAMOND GYPSUM MINE

The Blue Diamond Mine, which closed in 2004, was one of the two major gypsum mines in Clark County (the other mine, PABCO Gypsum, is still in operation and is seen on Trip 3). The 2,400-acre mine property was purchased by a Las Vegas developer who has hopes of building homes on the site. The adjacent Blue Diamond gypsum processing plant is still operated by BPB, a British company which is the largest producer of plasterboard in the world. Currently, however, gypsum that provides the raw material for this plant is shipped here from mines in Arizona. The plant produces plasterboard for the construction market in Las Vegas and nationwide.

At the Blue Diamond Mine, gypsum was extracted from a layer in the upper part of the Permian Kaibab Formation. Gypsum, which is hydrated calcium sulfate (a mineral composed of calcium sulfate with water included as part of its chemical makeup), forms deposits in warm, shallow water in arid environments, in this case most likely along the shores of the Permian ocean. As now exposed, the gypsum layer is on the gently westward sloping top of Blue Diamond Hill. Ore was mined from open pits along the hilltop, trucked to an inclined conveyor belt that carried the ore down the steep face of the hill into the processing plant. The conveyor belt is housed in what appears at a distance to be a large metal tube descending from the hilltop (still present in 2007).

What does the future hold for this mine which still contains a major gypsum resource? If the original plans of the developer who bought Blue Diamond Hill and the mine had come to pass, some 5,500 homes would have eventually graced the top of the hill. Fearing that this urban encroachment would impact the adjacent Red Rock Canyon National Recreation Area, zoning legislation was passed restricting development on private lands surrounding the recreation area. The new zoning, however, only restricts but does not stop development and someday, instead of one of Nevada's largest gypsum mines, there may be expensive homes with spectacular views of Red Rock Canyon rising from reclaimed gypsum mine dumps.

Photo: Bill Rogers

Conveyor that carried gypsum ore from the Blue Diamond Mine at the top of the cliff to the processing plant at the base of the cliff.

Photo: Joe Tingley

BPB (formerly James Hardie) Blue Diamond gypsum processing plant at the base of Blue Diamond Hill.

interval	cumulative	
0.4	12.2	At the narrows in the road, the ridge-forming rock is limestone of the Kaibab Formation, which is easily recognized by conspicuous brown concretions of chert. These chert concretions (chert is a form of quartz) are harder than the enclosing limestone, so they tend to stand out in relief as the surrounding rock weathers away. Concretions usually form at the same time as the enclosing sediments by the accumulation of mineral material around a nucleus, such as a plant or shell fragment, prior to the hardening or consolidation of the sediments into rock. The Kaibab Formation is the rock layer that forms the rim of the Grand Canyon. Fossil evidence indicates that it was deposited about 250 million years ago in a warm shallow sea similar to the environment that exists today near the Bahamas. The widespread distribution of the Kaibab Formation indicates the great extent of this former shallow sea. Although today they do not have a seaside view, barrel cactus seem to especially like the rock face to the right. *(GPS 11)*
0.3	12.5	Entering Red Rock Canyon National Conservation Area administered by the U.S. Department of the Interior's Bureau of Land Management (BLM). All items within the area are protected, so enjoy the area with your eyes but do not remove any rocks, plants, or other natural items. *(GPS 12)*
0.2	12.7	Town of Blue Diamond is on the left. Formerly a company town for workers at the nearby gypsum mine, the name comes from the Blue Diamond Corp. which began operations here in 1923. The striped-appearing rock outcrops to the west above town are the Moenkopi Formation. The more rugged rock layer outcropping on the skyline at 9:00 (above the two white water tanks) is a late Miocene (probably formed 17 to 10 million years ago) breccia, thought to be remnants of a huge landslide that broke away and slid from the Red Rock escarpment to the west.

Trees, Brush, and Other Things That Cover the Rocks

The route followed by this trip within the Red Rock Canyon National Conservation Area is in the Lower Sonoran life zone, known as the Mojave Desert. This part of the conservation area hosts a large variety of plant life, including various types of yucca, Joshua tree yucca (an indicator plant of the Mojave Desert), blackbrush, creosote bush, mesquite, catclaw, cholla, and many wildflowers. Near the base of the western escarpment, areas along streams and springs (riparian zones) support scrub oak, desert willow, redbud, Fremont cottonwood, and ash trees. The higher cliffs support vegetation of the Transition life zone including piñon, juniper, and ponderosa pine. More than 100 species of birds, 45 species of mammals, and 30 species of reptiles and amphibians, including the desert tortoise, live in the conservation area.

The **Joshua tree** (*Yucca brevifolia*) is a characteristic plant of the Mojave Desert. The largest of the yuccas, it can grow to heights of 30 feet or more. It can reproduce itself by sending up shoots from roots and rhizomes. In order to produce seeds, it carries on a symbiotic relationship with the small white **yucca moth** (*Tegeticula synethetica*), which in turn depends on the Joshua tree's seeds to feed its young. The yucca moth picks up and distributes pollen as it moves from flower to flower to deposit its eggs. Larvae develop within the fruit, feeding on a small portion of its seeds. Other species of yucca maintain similar relationships with other species of yucca moths.

Photo: Joe Tingley

interval cumulative

0.4 | 13.1 — The road into the town of Blue Diamond is on the left. There is a nice grassy park and picnic area if you want to stop for refreshment before beginning your tour of Red Rock Canyon. *(GPS 13)*

0.6 | 13.7 — A good view ahead of the red sandstone escarpment topped by the Wilson Cliffs thrust. You are looking directly up First Creek Canyon. Mount Wilson is the high peak to the right. The Wilson Cliffs thrust separates the gray limestone seen at the head of First Creek Canyon from the red and white Aztec Sandstone of the escarpment. *(GPS 14)*

0.3 | 14.0 — To the right, the gentle dip slope is capped by limestone of the Kaibab Formation. The valley ahead is formed in easily eroded siltstone and shale of the Moenkopi Formation. Beyond the curve ahead, still to the right, the gently inclined slope on the east side of the valley is composed of the Virgin Limestone Member of the Moenkopi Formation.

0.6 | 14.6 — Here the road passes through a rather nice Joshua tree forest, a sure sign that this is the Mojave Desert.

0.4 | 15.0 — Turnoff to Bonnie Springs Ranch on the left. It is the only commercial private land within the BLM's Red Rock Canyon National Conservation Area. *(GPS 16)*

Wild burros are commonly seen along this route. They are not indigenous to this area, but are descendants of domesticated burros that were abandoned or escaped from their owners during the early mining days. Protected by law since 1971, the burros are now under the jurisdiction of the BLM. Please do not feed the burros, there is a fine—and they bite!

Photo: Joe Tingley

Entrance to Bonnie Springs Ranch.

SPRINGS

Many springs and seeps occur at the base of the red sandstone escarpment. These issue water most voluminously during the spring and early summer, and a few springs run year round. The springs are located at the base of the cliffs because water and snowmelt that accumulate at higher elevations percolate downward into and through the permeable Aztec Sandstone (there is pore space between the sand grains that allows water to move through it, i.e., it's "permeable"). The rock layers that lie beneath the Aztec Sandstone belong to the Triassic Chinle Formation, which is composed of shale and siltstone. The pore spaces within the siltstone and shale are so small that the Chinle Formation is far less permeable than the overlying Aztec Sandstone. Water travels easily through the Aztec Sandstone, but cannot easily continue downward through the relatively impermeable Chinle Formation. Instead, the water discharges as springs at the face of the cliffs at low points along the contact of the two formations.

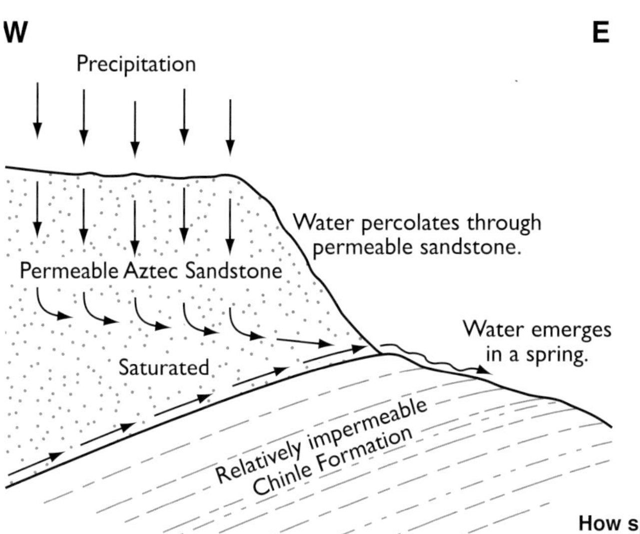

How springs form at the base of the Sandstone Bluffs.

Photo: Becky Purkey

A typical canyon cut into the east face of the Sandstone Bluffs. Vegetation at the canyon mouth is kept lush by springs.

interval	cumulative	
0.8	15.8	The turnoff to Spring Mountain Ranch State Park is on the left. This area has a long history of use. An established campsite was reported by early travelers in the mid-1830s. It was an alternate route for pack and wagon trains on the Old Spanish Trail, which passed through Cottonwood Valley to the south (now S.R. 160). Use of this trail ceased with the establishment of the San Pedro, Los Angeles and Salt Lake Railroad in Las Vegas Valley in 1905. The ranch has had an interesting string of owners who used it as both a working ranch and a luxurious retreat. It was purchased by the Nevada Division of State Parks in 1974 and is now open to the public. The ranch house can be toured, accompanied by very knowledgeable docents, and there are picnic sites, drinking water, and rest rooms. A day-use fee is charged. *(GPS 17)*
0.6	16.4	First Creek Trailhead on the left. *(GPS 18)*
0.6	17.0	The turnoff to Oak Creek is on the left. *(GPS 19)*
0.9	17.9	Good view of the Wilson Cliffs thrust ahead. The gray rocks on the horizon are Bonanza King limestones that form the upper plate of the thrust. The red and white Aztec Sandstone is in the lower plate. The thrust contact is to the left, behind the Red Rock escarpment. The contact then swings around to the east and crosses the valley ahead, behind the two points of red rock.
0.5	18.4	Passing the end of the Red Rock Scenic Drive on the left. The colorful, aptly named, Calico Hills are straight ahead.
0.5	18.9	Red Rock Overlook and interpretive display is on the left. Take the time to pull into the parking area here and check out the display. You can pick out all of the prominent topographic features and learn their names. This will make the rest of your trip more meaningful. There are rest rooms here, and also picnic tables (some are shaded). *(GPS 20)*
0.3	19.2	The private road to the right leads into the Blue Diamond gypsum mine. The mine dumps are visible on the skyline at the head of the canyon. *(GPS 21)*
1.5	20.7	Turn left into the BLM Red Rock Canyon National Conservation Area (there is a $5 fee, but they honor Golden Age Passports). Entrance into this area provides access to a 13-mile scenic drive, more than 30 miles of hiking trails, and several picnic areas (drinking water is available only at the Visitor Center). *(GPS 22)*
		While you are here, make sure you spend some time in the Visitor Center where you can check out the fascinating and informative displays on the natural and human history of this area—and don't overlook the well-stocked bookstore. This is the place to inquire about hiking trails and pick up trail maps, and you may also want to inquire about ranger-led interpretive tours in the area. Also here are a picnic area, water, and rest rooms.
		Leave the Visitor Center and backtrack to the entrance of the Red Rock Scenic Drive.

Artwork: Ralph Bennett

Photo: Carol McKim

The smallest of foxes, the **desert kit fox** (*Vulpes macrotis*) is about 30 inches long from nose to tail tip. It has very large ears and is gray and buff colored with a black-tipped tail. The bottoms of its feet are covered with fur to facilitate travel on sandy soil. Mostly nocturnal, the fox hunts rodents, lizards, birds, and insects.

Photo: Becky Purkey

Red Rock Canyon National Conservation Area Visitor Center.

Aztec Sandstone, the Rock Behind Red Rock Canyon

If we had to pick one unique feature to describe the Red Rock Canyon area, it would be this rock—the Aztec Sandstone (see color photos on front cover). For a rock, it has all the features that make it stand out from the ordinary. It has color—not uniform but in shades of red, white, tan, gray, and even purple—that occurs in patches and swirls governed by the vagaries of differential weathering and cross-bedding. In places, it is well cemented by calcite or silica into a fairly hard, competent rock, meaning it weathers into spectacular steep cliff faces. In other places where the cementing is variable and not so good, this rock weathers into lots of interesting forms with grottos and knobs (features well exposed in the Valley of Fire on Trip 3). Even its presence here is somewhat outstanding. We are used to seeing colorful sandstone to the east in the Grand Canyon, but here, among our normal Nevada rocks, these fiery red sandstone outcrops are striking.

This sandstone formation is actually present throughout much of the southwestern United States. In Zion National Park and the Lake Powell area of Utah and Arizona, this rock layer is called the Navajo Sandstone. In parts of Colorado and Utah, it is called the Nugget Formation. Our Nevada outcrops of Aztec Sandstone represent the westernmost extent of this widespread rock unit. In addition to the spectacular cliff-face exposures that you see on this trip, you will see more Aztec Sandstone in the Valley of Fire on Trip 3.

The sandstone is composed of well-rounded grains of almost pure quartz sand with no clay or silt. This indicates that the sand was deposited by some agent that was capable of removing fine material from the sediment and one that was efficient at rounding the edges of the sand grains by abrasion. Another diagnostic feature of the Aztec Sandstone is its conspicuous cross stratification or cross-bedding. Unlike most sedimentary deposits that show planar stratification, the sandstone cliffs show multiple thin, parallel to gently curving layers that are oriented at angles to one another. This type of stratification is a result of deposition by wind or water currents that were variable in direction and intensity. The large scale of individual cross-beds in the Aztec Sandstone is most characteristic of deposition by wind currents.

The widespread extent of the Aztec Sandstone and its correlative formations throughout much of the southwestern United States suggests that this formation was deposited in a vast desert sand dune environment much like the present-day Sahara Desert in north Africa.

Note that the sandstone varies in color from white or light tan to red and that these color changes do not coincide with specific layers or beds within the sandstone, but appear to occur independently of layering. The ultimate origin of these color bands is not fully understood, but they were probably caused by the movement of groundwater through the rock after it was formed. Upon prolonged exposure to oxygen and water in the subsurface, minerals that contain even small amounts of iron undergo oxidation, the same process that rusts metal. The result of this process is a red discoloration of the rock. Apparently, the lighter-colored areas somehow escaped oxidation; perhaps groundwater moved through the subsurface in an irregular fashion.

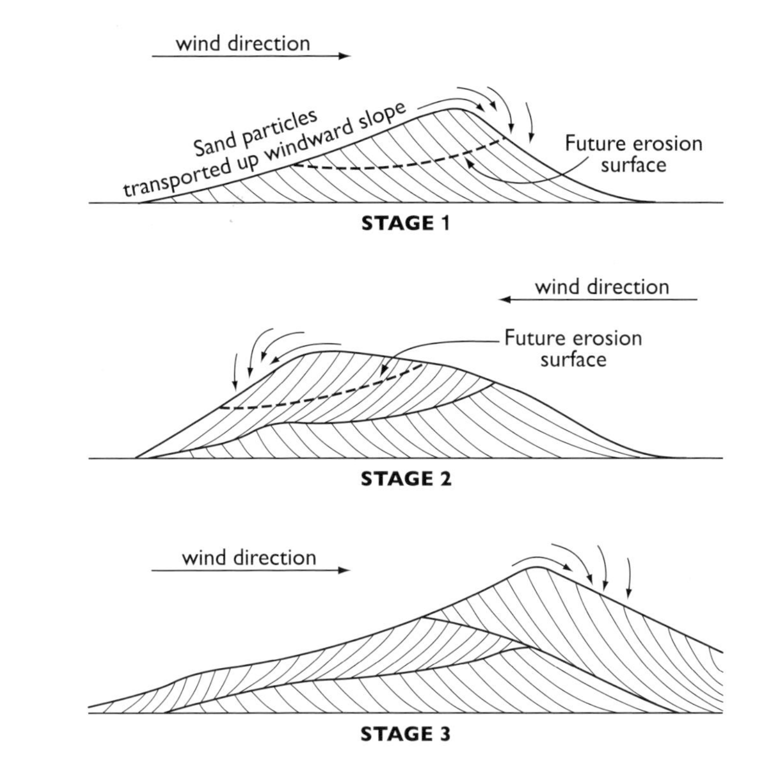

Formation of cross-beds in sandstone.

Cross-beds in Aztec Sandstone, Calico Hills.

interval | cumulative

BEGIN 13-MILE RED ROCK SCENIC DRIVE

0.0 — Reset odometer to 0.0, turn left, and begin the 13-mile, one-way scenic drive through Red Rock (the mileages given include entering turnouts, parking areas, etc. where noted). Between here and the two Calico Hills Overlooks, you will be driving through lands scarred by the Loop Fire which burned 869 acres in the eastern part of the Recreation Area in 2006. *(GPS 23)*

1.0 | 1.0 — Calico Hills No. 1 Overlook on the right. Stop for a close-up view of the Aztec Sandstone. This is a great spot to observe cross-bedding in the sandstone exposures. *(GPS 24)*

0.5 | 1.5 — Calico Hills No. 2 Overlook on the right. There are well-written interpretative signs here to check out. *(GPS 25)*

0.9 | 2.4 — Sandstone Quarry to the right. Enter the parking area. One of the first industries established in the Las Vegas area after 1905 was this quarry, which produced very hard (due to the presence of calcite or quartz cement between the sand grains) red and white sandstone for buildings in Las Vegas, Los Angeles, and San Francisco. A huge steam traction engine hauled the cut blocks to the railroad in Las Vegas (see sketch on lower left, page 22). The quarry closed in 1912 due to competition from a new quarry closer to the railroad. *(GPS 26)*

Photo: Becky Purkey

Aztec Sandstone displaying cross-beds and color variation.

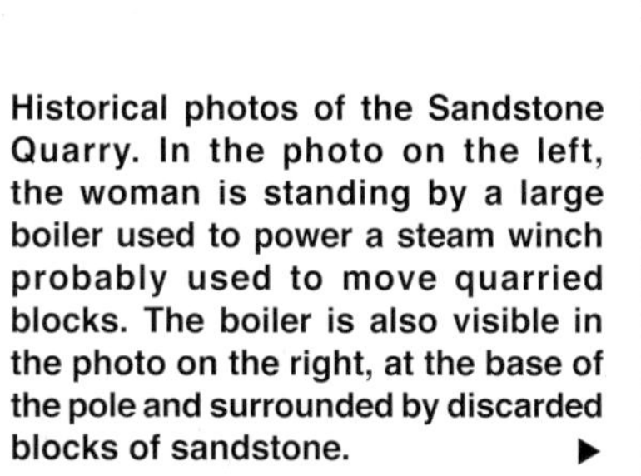

Photos: Joe Tingley

Sandstone Quarry, upper photo, and large quarried blocks that remain on site (lower photo).

Historical photos of the Sandstone Quarry. In the photo on the left, the woman is standing by a large boiler used to power a steam winch probably used to move quarried blocks. The boiler is also visible in the photo on the right, at the base of the pole and surrounded by discarded blocks of sandstone.

Photos: Red Rock Canyon National Conservation Area

interval	cumulative	
2.3	4.7	High Point Overlook on the left. From the observation area here, a view to the south across Red Rock Wash reveals the gentle westward inclination of the rock strata on both sides of the valley. The valley developed by preferential erosion of the soft shale and siltstone of the underlying Chinle and Moenkopi Formations. The escarpment to the west rises above the valley floor because the Aztec Sandstone is far more resistant to erosion than the softer layers beneath it. *(GPS 27)* There is a good view of the Blue Diamond gypsum mine on the top of Blue Diamond Hill, east of the valley. The road cuts on the right (north) side of the road show alluvial material eroded from the high cliffs of La Madre Mountain to the north. Near the surface, this alluvium has been partially cemented by a white, impervious material called caliche. (Note the thin white lenses in the alluvium and the white coatings on pebbles.) Caliche is composed mostly of calcium carbonate that has been dissolved from the limestone and dolomite of the higher parts of the Spring Mountains to the west, transported in surface and groundwater, and deposited (precipitated or crystallized) at depths of a few inches to several feet in pores in the soil. Caliche also forms coatings on cobbles resting on the surface by capillary rise of calcium-rich groundwater. There is not enough precipitation in arid climates to wash the calcium minerals away. Thick, hard caliche deposits pose numerous problems for developers in arid regions. (Refer to page 76 for more information on caliche.)
1.1	5.8	Turn off to White Rock Spring on the right. Drive up the road ½ mile to the parking area at the trailhead to White Rock Spring. As you enter the parking area, you will notice that the hill ahead of you and to your left is composed of Aztec Sandstone, but the sandstone outcrop is abruptly terminated at the wash to the right of the hill. A short distance to the east (your right from the parking lot) you can see a small hill of gray rock rising above the alluvium. What has happened here is a northwest-oriented, steeply inclined or nearly vertical fault named the La Madre fault (it roughly parallels and is a little east of the wash) has cut and offset the rocks. The block east of the fault has been dropped down relative to the block to the west, bringing the gray limestone of the older Bonanza King Formation (that lies above the Keystone thrust fault) down against the younger, tan Aztec Sandstone. Unfortunately you cannot see the fault contact because it is covered by alluvial material flooded down the wash. If you want to see upper plate rocks, take the trail up the wash for about ¼ mile, then take the trail to the right marked Keystone thrust. Not only will you see the upper plate limestone, you will see the thrust itself and the sandstone below the thrust exposed in a wash. This is a moderately strenuous hike of about 2¼ miles roundtrip. *(GPS 28)*

Photo: Joe Tingley

View to the south from High Point Overlook. Red Rock Wash crosses from right to left across the photo in the middle distance. The Sandstone Bluffs, with Mount Wilson on the right, are south of Red Rock Wash.

Photo: Becky Purkey

Caliche deposits exposed in a road cut near High Point Overlook.

interval	cumulative	
		To get a good look at the Aztec Sandstone in the outcrop west of the La Madre fault, take the fork to the left at the Keystone thrust trail intersection and continue northwest on the White Rock Spring/La Madre Spring Loop Trail. On this trail you will get a good look at the Aztec Sandstone folded into an overturned syncline as a result of the thrust faulting. The trail leads to the Willow Spring Picnic Area (described ahead at mileage 8.4). At Willow Spring, you will be looking at the west end of the same syncline, but the rock structure is easier to see here on the White Rock Spring end of the loop trail. If there are several in your party and not all want to hike, send the car ahead to Willow Spring to wait while the hikers take the trail. The hike is about 3 miles one way and is moderately strenuous.
1.6	7.4	More caliche deposits can be seen in the road cuts to the right.
1.0	8.4	Turn right to the Willow Spring Picnic Area (see color photo 3c on page 15). Drive to the picnic area and park. Although the road is passable to high-axle or four-wheel drive vehicles for a short distance beyond the picnic area, further travel is not recommended except on foot. *(GPS 29)*

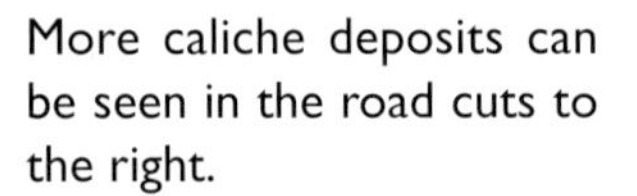

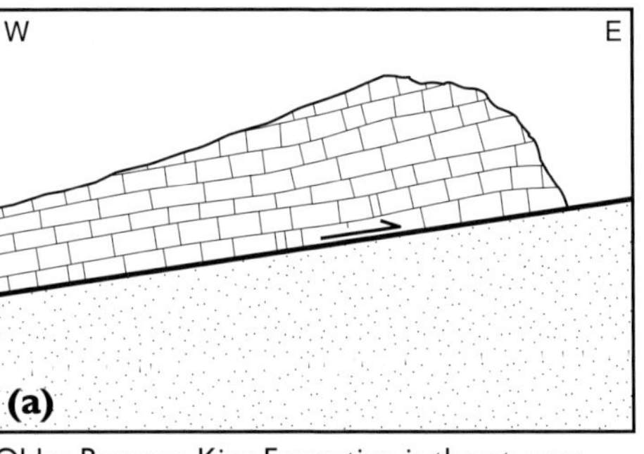

Older Bonanza King Formation is thrust over younger Aztec Sandstone.

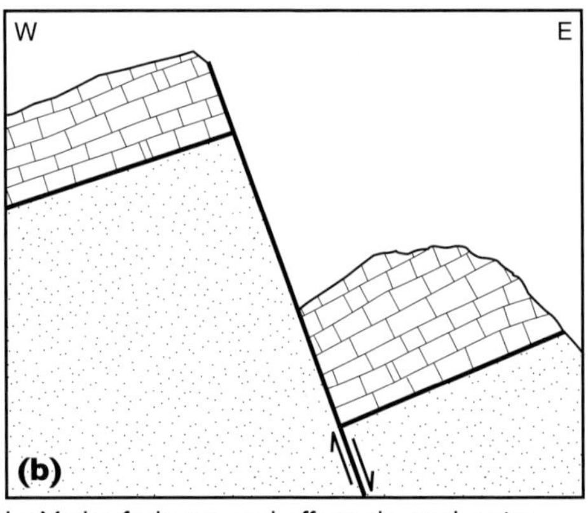

La Madre fault cuts and offsets the rock units.

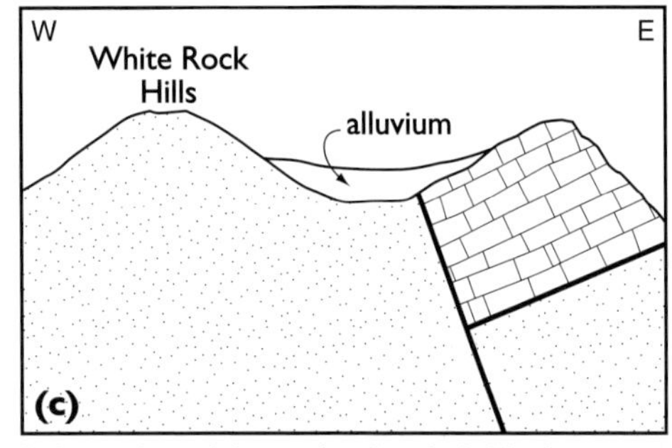

Relationship of the rocks after fault-block movement and erosion

approximately 1/2 mile

Geologic relationships at White Rock Spring. Sketches show development of offset across the La Madre fault.

Photo: Nevada Department of Highways

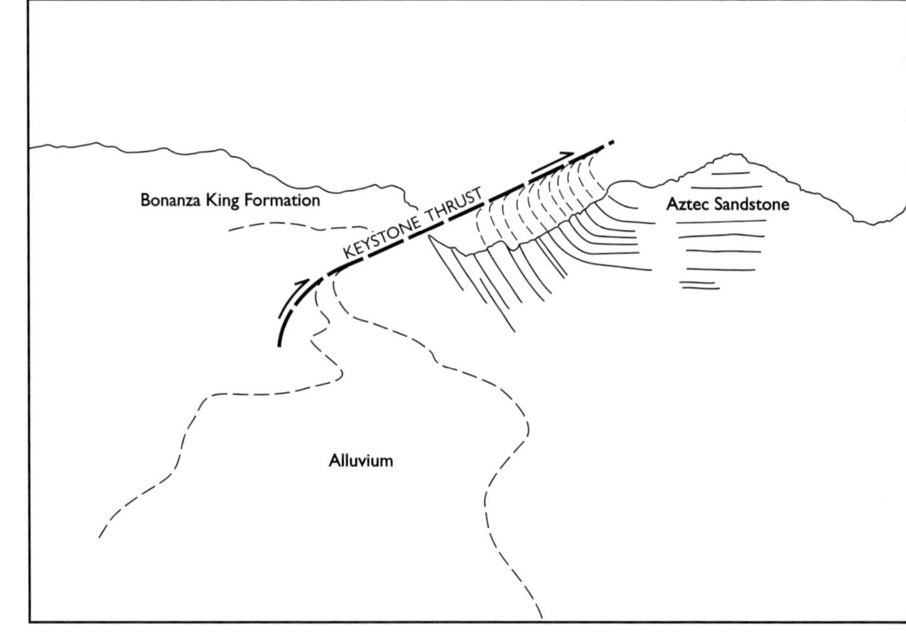

Photo and sketch of the north end of the White Rock Hills, view to the east from Red Roc Canyon. Layered rocks in the background to the left are limestone and dolomite of th Cambrian Bonanza King Formation in the upper plate of the Keystone thrust. Rugged outcrop in the White Rock Hills on the right are Aztec Sandstone in the lower plate of the thrust. A emphasized in the sketch, bedding layers in the sandstone were folded up then overturne as the thrust sheet was forced over the lower plate rocks.

interval cumulative

An optional hike west up the dirt road reveals interesting relationships beneath the Keystone thrust fault. Approximately ½ mile from the parking area, just beyond where the road crosses the streambed, look to the right. The Aztec Sandstone is inclined very steeply downward toward the valley to the southeast. This orientation is opposite to the gentle west inclination of the strata back at the parking area. The Aztec Sandstone has been folded into a bowl-shaped configuration called a syncline (see figure, page 30), and the syncline has been overturned (you would have seen the northeast side of this syncline from the White Rock Spring Trail, last stop, if you hiked from there).

Continuing up the road (keep left at the La Madre Springs Trail sign), the Aztec Sandstone ends, and road cuts now reveal dark red-brown siltstone and shale of the Chinle Formation that lies beneath the Aztec Sandstone. These layers have been tilted into a nearly vertical orientation. Continuing northwestward, the Keystone thrust fault contact is encountered at the base of the prominent gray cliffs. The cliffs are composed of gray limestone and dolomite of the Bonanza King Formation which, at this location, have been thrust over the younger Chinle Formation. Fossilized coral may be observed in the gray limestones and dolomites in the stream beds.

This area also has numerous archeological sites. Along the cliffs are petroglyphs and agave roasting pits attributed to the Ancient Puebloan culture (500 to 1150 A.D.) and southern Paiute Indians (1150 A.D. to recent time).

1.2 | 9.6 Return to the main loop road, turn right, and continue on. Note the dark patches of rock varnish coating the sandstone cliffs in this area. See page 85 for an explanation of rock varnish.

0.5 | 10.1 Ice Box Canyon Trailhead on the right. *(GPS 30)*

0.6 | 10.7 Crossing Red Rock Wash. *(GPS 31)*

0.3 | 11.0 Red Rock Wash Overlook on the left. This is a good place to pull in, stop, and enjoy the panoramic view. As you stand looking north over the wash, the Calico Hills, with their outcrops of lower plate Aztec Sandstone, are to the northeast at about 1:00 to 3:00. La Madre Mountain, composed of upper plate limestone, rises above the basin to the north. The White Rock Hills, composed of Aztec Sandstone, are to the northwest at about 11:00. *(GPS 32)*

At the overlook, you are standing on a large alluvial fan that was deposited at the mouth of Red Rock Canyon and is now being cut through and washed away by the modern intermittent stream that occupies the drainage. You can see the older cemented gravels of the fan exposed in the bank of the wash.

Beyond the overlook, the scenic drive passes through evidence of the Pine Creek Fire, one of some three dozen fires that burned about 2,300 acres in the Red Rock Canyon National Recreation Area in 2006.

Photo: Dr. Lloyd Glenn Ingles, California Academy of Sciences

Coyotes (*Canis latrans*) are grizzled gray or reddish-gray with buff underparts, a bushy tail with black tip, and prominent ears. They are excellent runners, with cruising speeds of 25 to 35 miles per hour and short bursts of up to 40 miles per hour.

Extremely intelligent and adaptable, the coyote is expanding its range despite loss of traditional habitat and human hunting pressures. Today the coyote can be found in desert, grassland, mountain, and suburban environments as far north as Alaska and as far south as Central America.

The coyote is an opportunistic hunter employing a variety of methods to obtain food. It patiently stalks and pounces on small mammals. It has tremendous endurance and can simply chase prey until it is worn out. Where the food supply is predominantly small animals, it hunts alone or in breeding pairs, while in the presence of large prey such as deer, it will hunt in packs. Coyotes will also make do with insects, lizards, carrion, fruit, and even pine nuts.

Coyotes are usually heard between dusk and dawn. Barks and yelps followed by a drawn-out howls serve to announce location, strengthen social bonds, and reunite separated members of a band.

interval	cumulative	
1.7	12.7	Pine Creek Canyon Overlook and Trailhead on the right. A fine example of Upper Sonoran vegetation may be observed here. Joshua trees and creosote bush of the Mojave Desert merge with piñon, ponderosa pine, juniper, aspen, and many species of shrubs. *(GPS 33)*
1.3	14.0	Oak Creek Trailhead to the right. *(GPS 34)*
0.6	14.6	End of Red Rock Scenic Drive. Turn left onto S.R. 159. *(GPS 35)*
2.3	16.9	Passing the entrance to the BLM Visitor Center. Return to Las Vegas via S.R. 159 and Charleston Blvd.
1.4	18.3	Calico Basin-Red Spring road on the left leads to the Red Spring Picnic Area (distance 1 mile). The major thrust fault here, called the Red Spring thrust for exposures near the picnic area, appears as a sharp line separating the gray limestone of the Cambrian Bonanza King Formation above, from the red Jurassic Aztec Sandstone below (see color photo 3d on page 15). *(GPS 36)*
0.5	18.8	Road to 13-Mile Campground on the right. *(GPS 37)*
0.8	19.6	Leaving Red Rock Canyon National Conservation Area. Note that the road now parallels Red Rock Wash on the left. This is a flash flood area. *(GPS 38)*
1.0	20.6	Red Rock basin flood detention structure on the right. The Red Rock basin accumulates snowmelt and runoff from the Sandstone Bluffs and La Madre Mountain. Spring snowmelt from these high areas is usually gradual enough that it does not normally constitute a flood hazard. However, summer thunderstorms that build up over the Spring Mountains can produce large amounts of runoff in very short periods of time. As in many parts of the desert, flash floods are common in southern Nevada; parking and camping in dry creek beds should be avoided. *(GPS 39)*

Photo: Mark Vollmer

Resident of Pine Creek.

Photo: Jack Hursh

Mormon tea *(Ephedra viridis)*. ▶

Photo: Joe Tingley

View to the east across Calico Basin from Red Spring. Gray rock above the dark band in the background is Cambrian limestone and dolomite in the upper plate of the Red Spring thrust (part of the Keystone thrust system). Rock below the dark band, and in the left foreground, is Aztec Sandstone in the lower plate of the thrust.

interval cumulative

Because the climb into the Red Rock area from Las Vegas is so gradual, it is easy to overlook the fact that the floor of the basin containing Red Rock Wash lies at an elevation of about 4,000 feet, nearly 2,000 feet higher than downtown Las Vegas. During heavy rains, runoff accumulates in the northern part of this basin and is channeled through Red Rock Wash into Las Vegas Valley. The Red Rock basin flood detention structure is designed to prevent floodwater from reaching the city of Las Vegas.

Las Vegas has a long history of major floods due to brief, but intense summer rains. Much of the city is built in low-lying areas such as washes or floodplains of washes. These are natural courses for runoff. The process of urbanization itself further exacerbates the flood hazard. Since paved surfaces are impermeable, water simply runs off and accumulates in the low places rather than infiltrating into the ground. Because of the relatively infrequent heavy rains in the Las Vegas area, city leaders and developers must make a choice between spending huge sums of money for flood detention structures before construction of buildings can begin, or opt for the lower costs of repairing damage after occasional flooding.

0.4 | 21.0 State Route 159 (Blue Diamond Rd.) becomes State Route 159 (Charleston Blvd.) *(GPS 40)*

1.0 | 22.0 Intersection, Charleston Blvd. and I-215. In the distance, to the east of the dense, urban development of the expanding city of Las Vegas, are Sunrise Mountain (lower peak at about 11:00), and Frenchman Mountain (higher peak at about 12:00). These are described in Trip 3. *(GPS 41)*

End of Trip 1.

Photo: Joe Tingley

Red Rock basin flood detention dam is in the background, beyond the highway bridge. The freshly smoothed gravel of Red Rock Wash in the foreground and under the bridge would look very different if flash-flood waters were moving down the wash.

Photo: Becky Purkey

Las Vegas and Frenchman Mountain, in 1993, viewed from Charleston Blvd. at Red Rock Wash. By 2007, this view had been obstructed by housing developments and large casino structures.

TRIP 2: THE SPRING MOUNTAINS

On this trip, we travel into the high central core of the Spring Mountains to view peaks and spectacular cliffs cut into Paleozoic limestone and dolomite. These rocks form the upper plate of the thrust faults that helped build this range of mountains in late Mesozoic time (about 100 to 90 million years ago). Trip 1, to Red Rock Canyon in the southeastern Spring Mountains, explores the red sandstones in the lower plate of this complicated thrust system.

In addition to geologic sight-seeing, the Spring Mountains offer both winter and summer recreation opportunities for dwellers of the southern Nevada desert. In the summer, temperatures are 20° to 25°F cooler than in Las Vegas, and the pine and fir forest provides a welcome relief from the heat for the hiker, picnicker, or cyclist. A small ski area, Lee Canyon, affords winter recreation less than an hour from downtown Las Vegas.

To reach the starting point of this trip, travel north from Las Vegas on U.S. Highway 95 (the road to Mercury and Tonopah). The trip begins at the intersection of U.S. Highway 95 and State Route 157. Turn left (west) on S.R. 157 to follow Kyle Canyon as it climbs into the Spring Mountains. From the end of S.R. 157, beneath Cathedral Rock high on the east flank of Charleston Peak, you can take advantage of many hiking trails to scenic points in the high country. Then backtrack a little, cross over to Lee Canyon via S.R. 158 and, after a visit to the Lee Canyon ski area, continue east on S.R. 156 down Lee Canyon and return to U.S. 95.

If taken in its entirety, this trip will cover nearly 100 miles. Be sure that you have enough gasoline before you start, because none is available in the Spring Mountains.

As you travel north toward the starting point, take note of the major intersections as you pass them. After passing the Santa Fe Station hotel and casino complex on the right at the junction of Rancho Drive with U.S. 95, there are major overcrossings at Ann Road, I-215, and Durango Road (Durango Road is the access to Floyd Lamb State Park at Tule Springs, discussed on Trip 5).

Continuing north on U.S. 95 beyond Durango, the Spring Mountains are to the left (at 9:00). The Sheep Range is at 1:00 to 1:30, and the southern tip of the Las Vegas Range, closest on the right, is at 2:00 to 3:00. The highest point in your view of the Las Vegas Range is Gass Peak (elevation 6,943 feet).

***Upper right photo:* Bristlecone pine branches frame a view of upper Kyle Canyon in the Spring Mountains.**

Limestone cliffs exposed in the upper reaches of Kyle Canyon in the Spring Mountains. Charleston Peak is on the skyline at the left and Coxcomb Ridge is at the right edge of the photo. The roads and houses visible along the canyon bottom mark the location of Mount Charleston (view is to the northwest from Cathedral Rock). ▶

Photos: Mark Vollmer

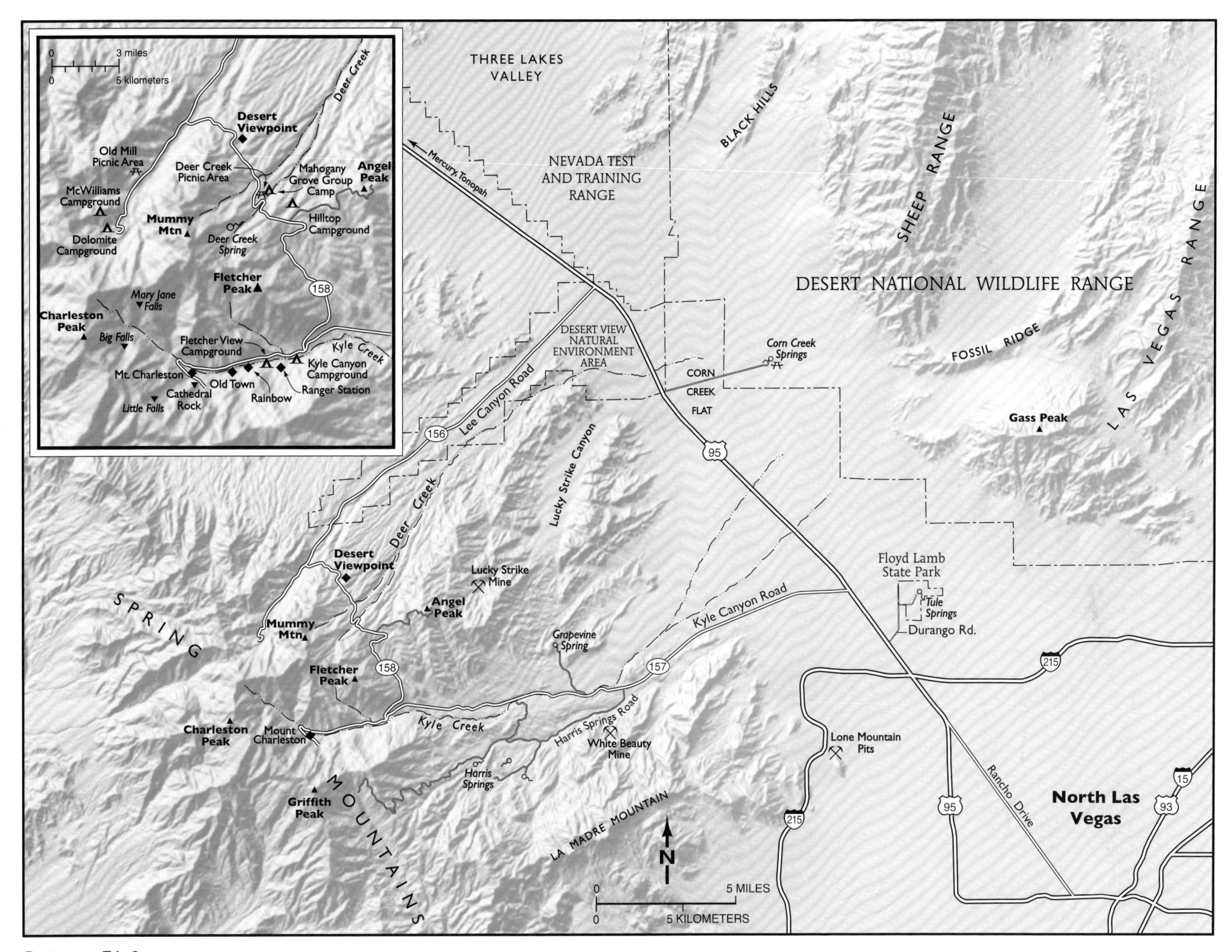

Route map, Trip 2.

interval / cumulative

La Madre Mountain is on the left, extending between 9:00 and 10:00. If the light is right (morning sun on the east-facing mountain), you can make out layered Paleozoic limestone and dolomite cropping out on the mountain face. The lower rocks are Ordovician (490 to 443 million years old) and Devonian (417 to 354 million years old) limestone and dolomite, and the main ridge is capped by Mississippian (354 to 323 million years old) limestone. The small outlier just north of the end of the main ridge is composed of Permian-Pennsylvanian Bird Spring Formation.

Turn left to Mount Charleston via Kyle Canyon Road (S.R. 157) and set your odometer to 0.0. *(GPS 42)*

— / 0.0 For the next 12 or so miles, the route is within the Lower Sonoran life zone, typical of most desert areas below about 3,500 feet above sea level. The scant 4 inches of average annual precipitation that falls in this area permits the growth of only the most hardy plants, those that are specially adapted to this arid climate, such as creosote bush, blackbrush, yucca, agave, mesquite, ephedra (Mormon tea), and various cactus species.

4.3 / 4.3 Ahead on the left, the smaller hill at 10:00 (this is the outlier mentioned earlier) is composed of well-stratified limestone and dolomite of the Permian-Pennsylvanian Bird Spring Formation. The larger hill at 9:00 is made up of Mississippian Monte Cristo Limestone. These latter rocks are equivalent in age to the prominent cliff-forming Redwall Limestone in the Grand Canyon. This limestone formed in shallow, warm oceans and contains fossils common to the Mississippian Period. The topographic slope is parallel to, and governed by, the slope (the dip) of the rock layers. A topographic slope that takes advantage of the underlying rock layers in this way is referred to by geologists as a dip slope.

2.7 / 7.0 For the next few miles, the route passes through a Joshua tree forest. Between about 3,500 and 5,500 feet above sea level, precipitation is sufficient to support the more varied vegetation of the Upper Sonoran life zone. This zone is typified by the dominance of shadscale, saltbush (in saline soils), and, here in the Mojave Desert, Joshua trees. Many plants from the Lower Sonoran life zone persist into the Upper Sonoran life zone. Rocks on both sides of the road are limestone and dolomite of the Bird Spring Formation.

Photo: Joe Tingley

Vegetation of the Lower Sonoran life zone along Kyle Canyon.

Photo: Becky Purkey

Vegetation of the Upper Sonoran life zone along Kyle Canyon.

Photo: Joe Tingley

Joshua tree forest. Joshua trees are members of the lily family and occur naturally only in the Mojave Desert of the southwestern United States.

Life Zones in the Spring Mountains

The different climatic zones encountered as one goes up in elevation or travels northward (in the northern hemisphere) result in notable differences in vegetation type. The 10,000-foot ascent from Las Vegas to the summit of Charleston Peak is equivalent, in terms of vegetation associations, to a trip from Las Vegas to the Canadian arctic and passes through six different life zones. We will pass through only four of these life zones on the trip up Kyle Canyon. You would need to hike to the top of Charleston Peak to encounter the upper two life zones.

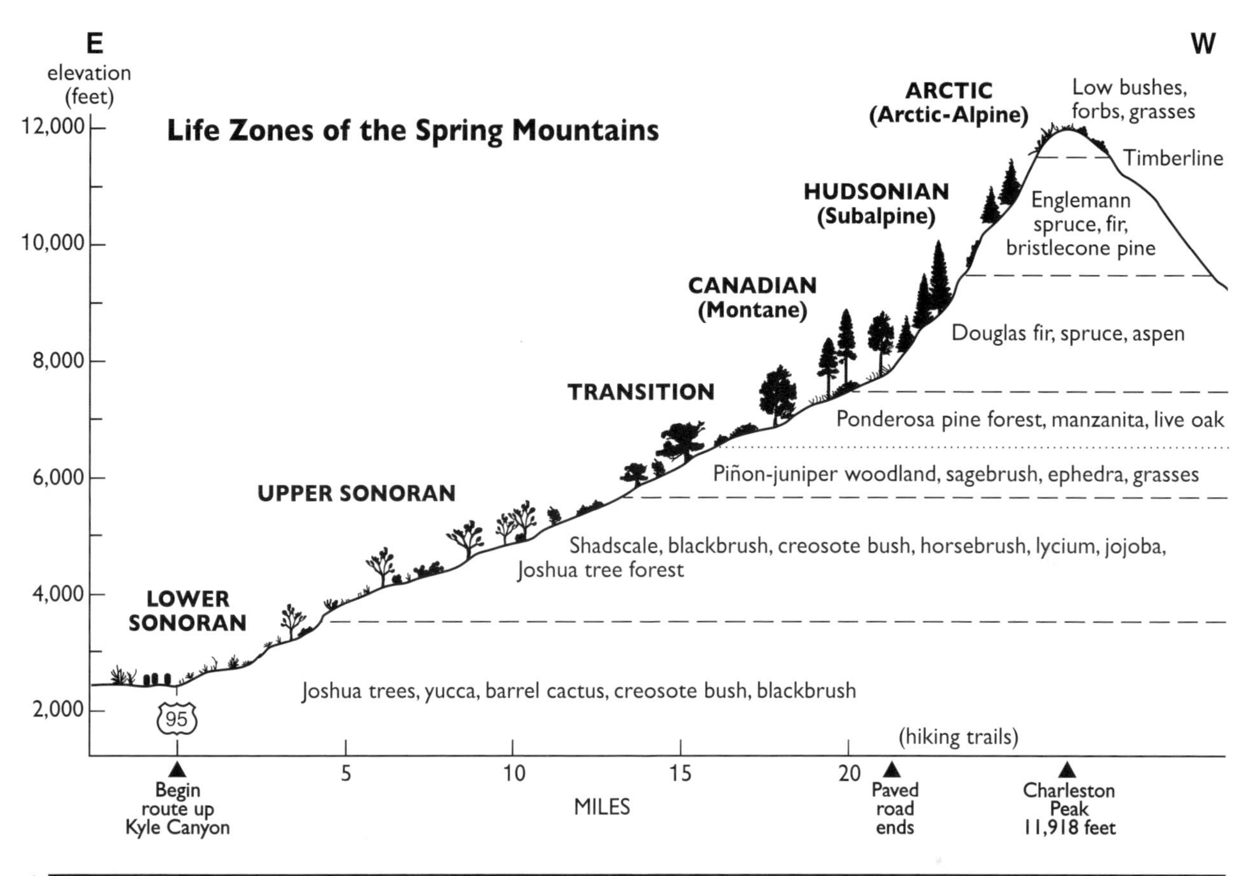

Line of bristlecone pine snags on the summit of Mummy Mountain (11,530 feet) northeast of Charleston Peak in the Spring Mountains.

interval	cumulative	
		The prominent white area at 8:00 to 9:00 high on the ridge is a relatively fresh landslide scar. During the landslide, a large slab of rock broke loose abruptly along a bedding surface that was a zone of weakness. This type of slope failure is referred to as a block slide. Gray, angular, limestone rubble—the actual landslide deposit—is evident below the landslide scar.
		This landslide serves as an example that erosion in steep mountainous areas is not always a slow and gradual process, but may be quite sudden and catastrophic.
1.0	8.0	The White Beauty gypsum mine is visible at the base of the cliff. A minor amount of gypsum has been produced here from small open pits in the Bird Spring Formation.
0.6	8.6	Road to the White Beauty Mine (Harris Springs Road) is to the left. The landslide scar is now seen at 9:00. *(GPS 43)*
0.4	9.0	The route now descends into Kyle Canyon wash. Deposits of the Pleistocene Kyle Canyon alluvial fan are evident above the road. The Kyle Canyon alluvial fan—actually composed of six distinct alluvial fan deposits—consists of large aprons of debris that were shed off the high Spring Mountains into stream drainages and deposited near the lower reaches of the range. The modern-day drainage system has cut down into these earlier alluvial fan deposits exposing their internal stratigraphy (layering). The large amount of debris in the Kyle Canyon alluvial fan dwarfs the volume of material that is presently being eroded from the Spring Mountains. This difference in intensity of erosion and the size of the resultant alluvial deposit is probably due to climate change. Analysis of fossil plant material indicates that precipitation in southern Nevada was much greater during the Pleistocene Epoch (Ice Age) than at present.
		The bouldery layers exposed in Las Vegas Wash at the Northshore Road bridge (see page 94) are probably the same age as the bouldery layers seen here in the Kyle Canyon fan.
0.8	9.8	On the right, tan Kyle Canyon fan material is plastered against gray limestone of the Bird Spring Formation. *(GPS 44)*
0.1	9.9	Folds within the Bird Spring Formation are evident to the left (south) of the road. At 9:00, the layers are inclined (dip) to the northeast. Ahead to the west (at 10:00 to 10:30), the layers dip northwest. This folding is related to a major phase of compressional deformation that occurred during the about 90 to 100 million years ago during the Cretaceous Period.

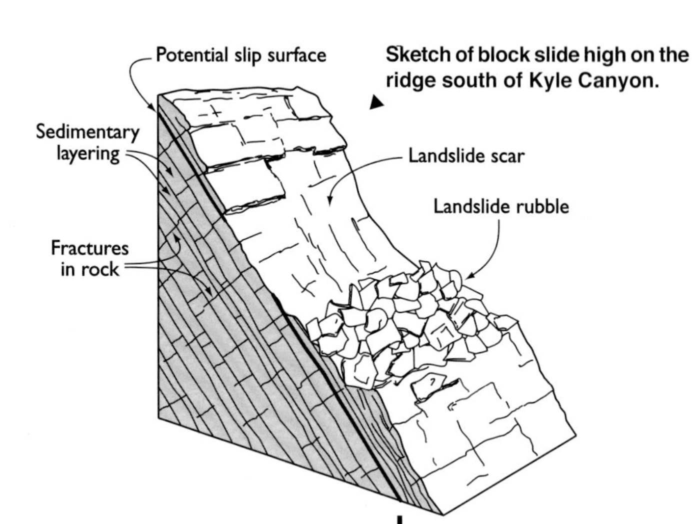

Sketch of block slide high on the ridge south of Kyle Canyon. ▶

Photo: Becky Purkey

Aerial view to the northwest across Kyle Canyon. Eroded deposits of the Kyle Canyon fan are in the center of the photo. Charleston Peak is in the upper left corner of the photo. ▶

Folds within the Bird Spring Formation exposed on the south side of Kyle Canyon. ▼

Photo: Joe Tingley

interval	cumulative
0.5	10.4

The crudely defined and irregular layering of the boulders, cobbles, pebbles, and sand displayed by this fan deposit is typical of material carried by debris flows or mudflows. These types of deposits involve minor amounts of water relative to the amount of sediment—just enough to move rock, sand, and mud material as a thick, sticky mass. In contrast, normal river deposits, which involve large amounts of water relative to the amount of sediment, show more defined and even stratification (layering) as well as more uniform size material. The larger, heavier pebbles settle to the bottom of the stream deposit first and the sediments get finer toward the top of the deposit because the finer sediments are lighter and are suspended longer in the water.

0.4 | 10.8

Isolated blocks and spires of gravel that you see on both sides of the road are parts of the ancient fan deposits that have not yet been eroded away by the present-day Kyle Creek. The older deposits are heavily cemented with calcium carbonate derived from the limestone rocks forming the bulk of the range. The resulting deposit, called calcrete, is similar to the caliche seen on Trips 1 and 3, but is formed in the presence of more moisture, such as the bottom of streambeds and lakes that cyclically fill and evaporate. These erosional remnants are held up by cemented, gravel-rich layers that are more resistant to erosion than the sediments that underlie them. Even more spectacular spires than these can be seen in Techatticup Wash between Nelson and Lake Mohave on Trip 4. *(GPS 45)*

1.5 | 12.3

Entering Spring Mountains National Recreation Area. Notice the gravel layers exposed in the recessed cliff face to the right. You can see a sequence of finer-grained, tan silty material and gravel exposed in the lower half of the cliff, capped by coarser, gray gravels above. This is good evidence that these gravels represent more than one period of deposition. *(GPS 46)*

0.3 | 12.6

The turnoff to Harris Springs Road is on the left. The route enters the Transition life zone which is marked by the rather abrupt appearance of piñon and juniper trees. This plant association (informally called PJ) is common throughout the southwest. In many areas, one plant species will dominate over the other; however, the two species are present in approximately equal amounts here. The piñons are the larger, dusty green trees with large needles and well-defined trunks. The junipers are brighter green and more shrublike. Other vegetation types include rabbitbrush, sagebrush, and wildflowers. Ponderosa pine, mountain mahogany, and oak dominate the highest elevations of the Transition life zone (6,500 to 7,000 feet). *(GPS 47)*

Photo: Joe Tingley

Blocks and spires sculpted by erosion in cemented gravel of the Pleistocene Kyle Canyon fan deposits.

Photo: Joe Tingley

Remnant of alluvial fan material plastered against bedrock exposed on the north side of Kyle Canyon.

interval	cumulative	
3.0	15.6	There is a good view of Charleston Peak at 11:30—the treeless peak on the right of the long ridge on the skyline. Fletcher Peak is at about 12:30. The road scar at its base marks S.R. 158, the road we will take after we finish exploring Kyle Canyon and Mount Charleston. *(GPS 48)* The prominent cliffs of Fletcher Peak are composed of Devonian and Mississippian dolomite and limestone situated in the upper plate of one of the regional thrust faults. At this location, these older rocks are thrust over Bird Spring Formation (most of the tree and shrub covered slopes on both sides of the highway are underlain by Bird Spring Formation). The La Madre fault, a major northwest-trending normal fault, passes across the road ahead, separating Fletcher Peak from the low hills to the right at 1:00 to 3:00. This fault extends southeast and cuts the ridge to the left at about 9:00. The southern extension of this fault is seen at White Rock Spring on Trip 1.
1.4	17.0	Entrance to the Charleston Hotel parking lot is on the left. Griffith Peak is the high point on the ridge straight ahead, and Fletcher Peak is above the road on the right at 3:00 to 4:00. Remnants of tan and gray, cemented Kyle Canyon fan gravel form cliffs behind the hotel (on the opposite side of the drainage). *(GPS 49)*
0.2	17.2	The active channel of Kyle Creek is on the left. Large boulders provide evidence of erosion and transport of coarse alluvial material during flooding. One reason that the creeks in the Spring Mountains flow so infrequently is that the limestone and dolomite that compose most of the range are highly fractured and contain numerous cavities formed due to dissolution (dissolving) of the limestone. Hence, most of the drainage is in the subsurface through these interconnected cavities. This relatively open area marks the intersection of at least three faults.
0.4	17.6	The Lee Canyon turnoff (S.R. 158) is on the right. Continue straight ahead through a ponderosa (yellow) pine and oak forest that marks the uppermost levels of the Transition life zone. Between this point and the Kyle Canyon Ranger Station ahead, the road is parallel to the trace of the Kyle Canyon thrust fault and crosses it twice. Rocks on both sides of the fault are Bird Spring Formation. *(GPS 50)*
0.4	18.0	Kyle Canyon Campground is on the left. Charleston Peak (11,918 feet) looms straight ahead. *(GPS 51)*
0.2	18.2	Entrance to Spring Mountains National Resource Area Visitor Center is on the left. The small building is stocked with books, maps, other informational material on the area, and a supply of tee shirts. There are also trail maps and exhibits on local history. *(GPS 52)*
0.2	18.4	Fletcher View Campground is on the left, Kyle RV Camp is ahead on the right. *(GPS 53)*

Photo: Joe Tingley

Piñon-juniper woodland of the Transition life zone. Piñons are larger, dusty green trees with large needles and well-defined trunks. The junipers are brighter green and more shrublike.

Photo: Becky Purkey

Ponderosa pine and oak forest characteristic of the upper levels of the Transition life zone.

interval cumulative

0.6 | 19.0 We now pass through the "town" of Rainbow. This is an informal name given to a cluster of cabins. No services are available. The elevation here is about 7,300 feet above sea level. Scattered aspen, Douglas fir, and spruce signal that the route has entered the Canadian (or Montane) life zone. This is the highest life zone encountered along the road. *(GPS 54)*

A hike to the summit of Charleston Peak via either the North Loop Trail or the South Loop Trail (consult U.S. Forest Service maps) will take you through the Hudsonian (or Subalpine) life zone dominated by Englemann spruce, and various species of fir. The Arctic (or Arctic-Alpine) life zone begins at timberline. It is equivalent to Arctic tundra and is characterized by lack of tree species and domination by low plants.

Note the outcrops of steeply tilted limestone in the road cut on the right.

1.0 | 20.0 Old Town. No services available. *(GPS 55)*

0.3 | 20.3 More Kyle Canyon fan deposits are on the right and, in the spectacular cliff faces above the road, there are steeply tilted exposures of Mississippian, Devonian, and Ordovician limestone and dolomite.

0.2 | 20.5 Hairpin turn left to Cathedral Rock, composed of massive Mississippian limestone. For access to the Trail Canyon and Mary Jane Falls Trailheads, continue straight ahead on Echo Road and follow the signs. At 0.2 mile, a gravel road leads to the left for another 0.4 mile to the Mary Jane Falls Trailhead parking or, by following the pavement to the right for another 0.1 mile, you arrive at the parking area for the Trail Canyon Trail.

The short trip to Mary Jane Falls Trailhead is worth it just to view the shear cliff face cut in Mississippian limestone above the parking area to the south.

0.4 | 20.9 Large paved trailhead parking area (with restrooms) for Little Falls Trail, Cathedral Rock Trail, and South Loop Trail. *(GPS 56)*

Maps and information regarding the trails are available from the Kyle Canyon Ranger Station (open between May 1 and September 30) or in Las Vegas. Hikes on this mountain should be undertaken only with proper Forest Service maps, adequate footwear, and a good supply of water.

Photos: Becky Purkey

Aspen, Douglas fir, and spruce characteristic of the Canadian (or Montane) life zone.

◀ View to the west from the parking area of Mount Charleston Lodge. The valley is bounded by high cliffs formed of Mississippian and Devonian limestone and dolomite, which are cut off by the nearly vertical Griffith fault.

Mount Charleston —A Populated Place

The U.S. Geological Survey topographic maps that cover this part of the Spring Mountains do not show locations named "Rainbow" or "Old Town," the informally named places you have just passed. Rather, the collection of cabins and summer homes in this entire section of Kyle Canyon lying below Cathedral Rock is shown on maps as "Mount Charleston" (category, "populated place"), not to be confused with Charleston Peak, the 11,918-foot topographic feature located a short distance to the northwest. The proximity of these two very different features with similar names brings up the importance of being precise when using place names.

interval	cumulative	
0.1	21.0	"Y" intersection in road. The road on the right leads to the Cathedral Rock Picnic Area (fee area). The left fork continues for another 0.2 mile where it ends at Mount Charleston Lodge. There is no parking on the one-way loop road above Mount Charleston Lodge. You can, however, pull into the Lodge parking area, turn around, and face west for a great view of the treeless summit of Charleston Peak. *(GPS 57)*

The valley to the west is bounded on each side by a line of high cliffs formed of Mississippian and Devonian limestone and dolomite. The nearly vertical Griffith fault passes through the valley generally along the base of the cliffs on the left. Ground breakage and offset along this structure has allowed the steep, cliff-bounded valley to form.

Alluvial fan deposits of the Kyle Canyon fan are exposed in the roadcuts to the Cathedral Rock Picnic Area, indicating that the present-day canyon was choked with fan debris at least this far up the mountain. The present-day drainage of Kyle Creek has cut through (dissected) these deposits.

Leave the Lodge parking area and begin your trip back down Kyle Canyon.

Below the Cathedral Rock Picnic Area, notice that the rocks on the north side (left after passing back through the hairpin turn) of the canyon are tilted in various directions. A series of faults associated with the Deer Creek thrust fault (to be observed later) cut across the road just east of the hairpin turn. Deformation due to movement along these faults caused the variable and chaotic tilting of the rock layers you now see here.

Continue down canyon and retrace your route to the intersection with S.R. 158.

interval	cumulative	
	0.0	Turn left (north) on S.R. 158 (Deer Creek Road) and reset odometer to 0.0. *(GPS 58)*
0.2	0.2	The cliff face straight ahead is composed of Mississippian Monte Cristo Limestone. This rock forms the upper plate of a thrust sheet. The less conspicuous lower plate rocks are marked by the brush and (sparse) tree-covered talus slope below the cliff.
0.8	1.0	As you pass through the thrust sheet below the Mississippian limestone, notice lots of broken, jumbled rocks.

La Madre Mountain is the long, low ridge on the skyline at 2:00 to 3:00. This view is toward the north side of the mountain. The rocks that make up the north side of La Madre Mountain lie in the upper plate of the Keystone thrust and have been folded into a large syncline (refer to the lower left figure on page 8). The core of the syncline is the Moenkopi Formation (red rocks). The winding route (S.R. 157, you drove it!) is constructed on the Bird Spring Formation.

Photo: Mark Vollmer

Mummy Mountain viewed from Lee Canyon.

Photos: Kris Pizarro

Mountain mahogany (*Cercocarpus ledifolius*) is found in the upper part of the Transition life zone and dense stands of it grow along Deer Creek Road (S.R. 158) between Kyle and Lee Canyons. Mountain mahogany is a small evergreen shrub or small tree with twisted branches that grow from a short trunk. The leaves are ½ to 1 inches long and inconspicuous funnel-shaped flowers are produced in spring. Seeds mature in late summer and are about ¼ inches long with a 1½ to 3-inch-long twisted tail covered with whitish hairs (*see inset*).

interval	cumulative	
0.9	1.9	Roadcuts expose more Kyle Canyon fan debris. The trace of the northwest-trending La Madre fault lies parallel to and just below road level on the right for about the next 2 miles. This is a normal fault, with the right side dropped down (see figure on page 8 for example of a normal fault).
1.2	3.1	On the north and east slopes (left side of road), vegetation zones are conspicuously mixed. Ponderosa pine, spruce, and a few aspen of the Transition and Canadian life zones occur in the protected and sheltered areas such as along streams. On more exposed slopes, piñon and juniper of the lower Transition life zone predominate, and there are even some magnificent stands of mountain mahogany along with serviceberry and elderberry bushes. The difference in vegetation types here is not due to elevation alone, but also to environmental factors such as degree of exposure to the sun, moisture retention properties of the soil, and depth of the soil. Much intermixing of vegetation types occurs in transitional areas.
0.3	3.4	Robbers Roost informational sign is on the right. There is a short trail that climbs up a canyon to the right of the highway to a couple of caves. The name springs from the popular legend that the caves offered a hideout for bandits who preyed on travelers following the old Mormon Trail between California and Utah. *(GPS 59)*
0.1	3.5	Roadcuts on both sides of the highway cut through deposits of the Kyle Canyon fan. Note that individual boulders, or clasts, locally exceed 2 feet in diameter. These large blocks reflect the proximity of the fan deposits to their source—the high Spring Mountains.
0.6	4.1	Angel Peak can be seen on the right, with relay towers and a white dome.
0.4	4.5	The turnoff to Hilltop Campground (SMYC sign) is on the right. The prominent mountain directly ahead (west) is Mummy Mountain, composed mostly of Mississippian and Devonian limestones. At an elevation of over 11,530 feet, it is not uncommon for snow to persist on this mountain until early June. Its profile, nearly 3 miles long, is very similar to a person lying down—hence the name. *(GPS 60)*
0.2	4.7	Notice the syncline in limestone exposed in the cliff to the left.
0.1	4.8	Trailhead, North Loop Trail to Charleston Peak. *(GPS 61)*
0.5	5.3	Mahogany Grove Group Camping Area on the right. *(GPS 62)*

◀ **Bristlecone snag on the summit of Mummy Mountain.**

◀ **Exposed bristlecone roots.**

Photos: Mark Vollmer

Mummy Mountain and vegetation characteristic of the Canadian life zone. ▼

interval	cumulative	
0.2	5.5	Deer Creek Picnic Area parking lot on the right (also rest rooms). Park, get out, stretch your legs, and examine the highly sheared limestones that have been caught up in the Deer Creek thrust fault zone. Movement along the Deer Creek thrust fault resulted in crushing and grinding of rock into angular fragments (fault breccia) and rock powder (fault gouge). This fault is part of the Keystone thrust system that is described in Trip 1. Although you cannot see the separate features at this location (all we see here are broken rocks in the wide fault zone), the fault places rocks of the Cambrian Bonanza King Formation over the younger Mississippian Monte Cristo Limestone. *(GPS 63)*

On the south (right) side of the road, the rocks have been completely overturned and overridden by the thrust plate.

On the north (left) side of the road, note the brecciated rocks that contain highly polished and reflective surfaces that have been produced by abrasive action during movement along the fault. Such polished and grooved surfaces are called slickensides. The orientation of the grooves on the slickenside surface provides geologists with important clues regarding the direction of movement (horizontal, vertical, or oblique) along the fault surface. It's easy to see which way the fault last moved, simply note the direction of the grooving. To tell which side of the fault moved, run your hands along the slickensides. If it feels smooth and your hand doesn't catch on little rough spots, it's moving the same direction the block of rock moved across them (which means the other block moved the other way, relatively).

interval	cumulative	
0.8	6.3	Dark gray Paleozoic limestone is exposed in the center of the roadcut on the left. Notice, however, that cemented alluvial fan gravels lap up onto each side of the limestone. This limestone is what we call a bedrock high—more simply, a hill—that was filled around by gravels washing from upstream. An irregular topographic surface existed at the time of deposition of the Kyle Canyon fan. As the fan debris was shed from the Spring Mountains, it first filled existing valleys and other low-lying areas. These valleys were separated by ridges or divides composed of the limestone bedrock. As deposition of the fan material continued, many of these divides were eventually buried beneath fan debris resulting in the type of situation you see in this roadcut. *(GPS 64)*

The former high-standing ridges and divides of limestone are evident in roadcuts where Paleozoic limestone is exposed. A modern-day example of this type of ridge and valley topography can be seen below the intersection of the Lee Canyon and Deer Creek roads.

Photo: Becky Purkey

Crushed rock in the Deer Creek thrust fault zone exposed in a roadcut north of the Deer Creek Picnic Area parking lot.

Photo: Becky Purkey

Slickensides (grooves gouged when fault surfaces move against each other) on a large rock fragment within the Deer Creek thrust fault zone.

interval	cumulative	
1.0	7.3	Desert Viewpoint on the right. The white color of the limestone at the far end of the large roadcut is due to bleaching related to intense faulting in this area. The crushed rock along the fault zone provides an easy path for percolation of surface water. This water, which here is slightly acidic, caused mineralogical and chemical changes to the crushed rock, resulting in the bleached appearance. *(GPS 65)* Stop here, walk the short, paved path to the overlook of Las Vegas Valley and the ranges beyond. From here, you are looking directly across the Las Vegas Valley shear zone, the structural footprint of basin-and-range faulting that has left such a major impression on this part of southern Nevada. Look at the panorama of mountains and valleys displayed on the plaque at the overlook, pick out the Desert Range and the small hills in the dry valley to its right and notice how they curve to your left as they approach Las Vegas Valley. These mountains have been bent or dragged to the left by movement along the Las Vegas Valley shear zone.

View to the northeast from the Desert Viewpoint. The extensive Desert National Wildlife Range includes much of the land in the background across the valley.

Las Vegas Valley Shear Zone

The Las Vegas Valley shear zone is one of the major structural features of this region. The shear zone is reflected topographically by Las Vegas Valley, and U.S. 95 follows the shear zone for about 55 miles from Las Vegas to the vicinity of Mercury, Nevada. There is thought to be 30 to 35 miles of right-lateral offset across this zone. This means that rocks that match those in the Las Vegas Range to the right are now to be found several tens of miles ahead, north of Lee Canyon in the Spring Mountains, on the left. The sharp bending, or drag into the shear zone of the axes of the major mountain ranges is easily seen evidence of this movement.

The age of the strike-slip movement is thought to be late Miocene (16 to about 5 million years ago), but movement could have extended to early Pliocene (less than 5.3 million years ago).

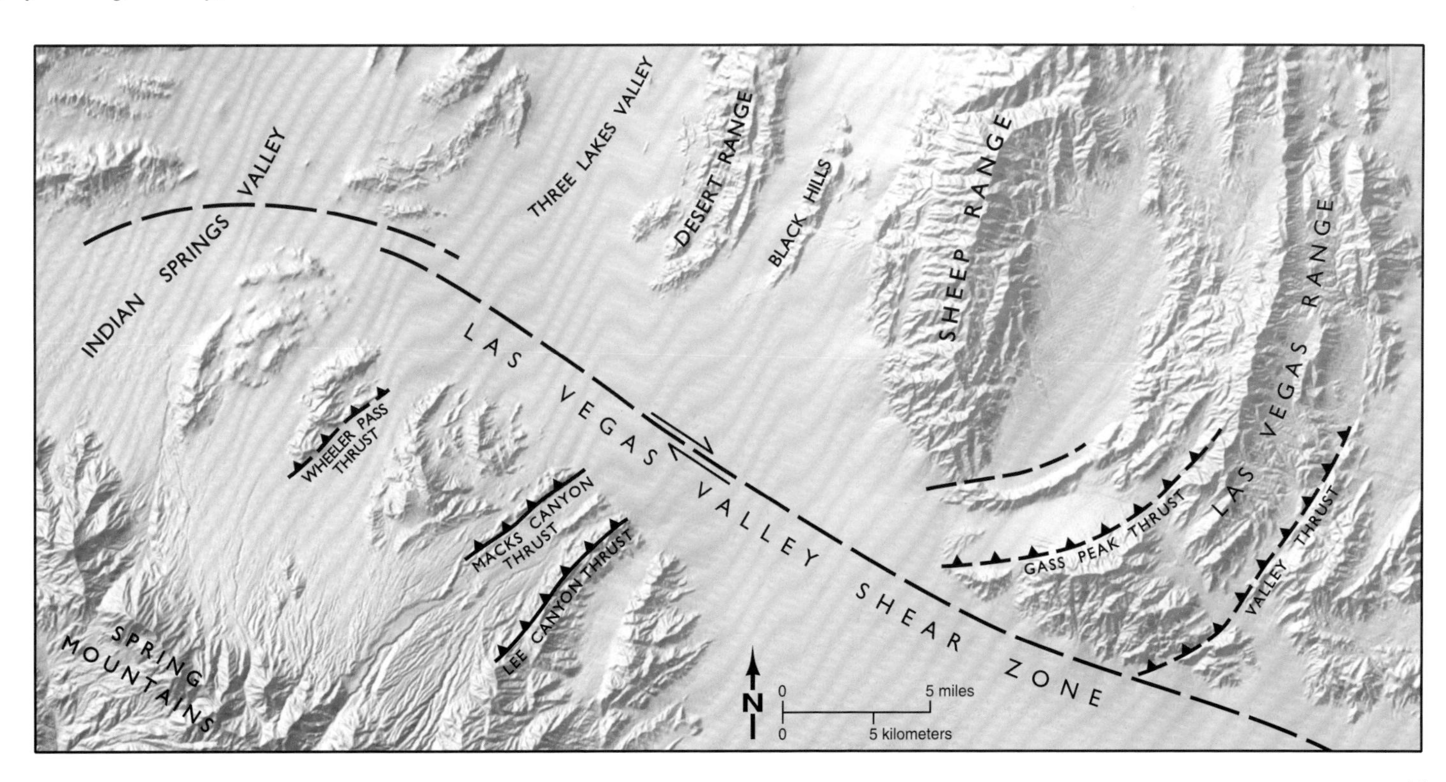

Trace of the Las Vegas Valley shear zone, upper Las Vegas Valley between the Spring Mountains (on the left) and the Las Vegas and Sheep Ranges (on the right).

interval	cumulative	
		Now look for Lucky Strike Canyon (it's about 5 miles due east, check the information on the plaque at the view point railing). There is a small red spot showing through the trees on the low ridge following the north side of the canyon. This is the site of the historical Lucky Strike lead mine. This area, known as Amber Mountain to the early Mormon miners, was the site of the first lead discovery in Nevada. The discovery was made in 1856 by a party from the Las Vegas Mission.
		Beyond this stop, the Desert View Natural Environment Area extends out on the dry alluvial fans to U.S. 95. The Desert National Wildlife Range extends along the east side of U.S. 95, encompassing six major mountain ranges including the Sheep and Las Vegas Ranges from Indian Springs on the north to Las Vegas on the south—almost 1.5 million acres. It was created in 1936 and is the largest wildlife refuge in the contiguous United States. Its most important objective is to perpetuate the desert bighorn sheep and its habitat, and it has the single largest concentration of these sheep in the country.
0.2	7.5	Mummy Mountain is straight ahead, on the far skyline.
0.4	7.9	The abundant mountain mahogany on both sides of the road here is growing on Ordovician limestone and dolomite.
1.0	8.9	Intersection with S.R. 156 (Lee Canyon Road). Turn left to Lee Canyon Ski Area. Lee Canyon is formed along the trace of a major thrust fault (the Lee Canyon thrust) for most of its length. Rocks to the north (right) are in the upper plate of this thrust, which here places Cambrian rocks (540 to 490 million years old) over Ordovician rocks (490 to 443 million years old). The Cambrian rocks form the high peaks. *(GPS 66)*

Photo: Becky Purkey

View to the south of a portion of the Lee Canyon Ski Area.

Desert bighorn sheep.

Artwork: Ralph Bennett

Photo: Joe Tingley

An old wooden headframe over a shaft at one of the small lead-zinc mines in Lucky Strike Canyon.

interval	cumulative	
1.8	10.7	Old Mill Picnic Area. *(GPS 67)*
0.2	10.9	Lee Canyon meadow. There is not another mountain meadow for more than 150 miles in any direction. Straight ahead, ski runs are visible on the mountain. *(GPS 68)*
		The high peaks to the left are composed of cliff-forming Mississippian limestone (354 to 323 million years old) underlain by Devonian rocks (417 to 354 million years old). The abrupt change, from prominent cliff on the left to more subdued outcrop on the right, is due to a fault offsetting the cliff-forming limestone beds. Relative to the rocks on the left (east of the fault), rocks to the right (west of the fault) have moved down; the cliff-forming rocks lie beneath the surface here.
0.6	11.5	McWilliams Campground to the right. Campgrounds in the Spring Mountains area are generally open from May 1 to September 1. This is a popular camping area and reservations may be required on weekends and holidays. *(GPS 69)*
0.2	11.7	Dolomite Campground to the right. *(GPS 70)*
0.6	12.3	The road ends at the turnaround for the Bristle Cone Trailhead at the upper end of the Lee Canyon Ski Area parking lot. Turn around here and begin your return descent of the canyon. *(GPS 71)*
		The trace of the Lee Canyon thrust fault more or less follows the embankment that extends from the helipad, back downhill to the point the road curves sharply into the upper parking area. The fault trace then gently curves to the right to follow the valley to the northwest. The crushed limestone in the embankment is in the fault zone and extends into the upper plate of the thrust. The massive cliffs to the east (directly ahead as you descend down canyon below the ski area) are Mississippian and Devonian limestone and dolomite.
		Continue down the mountain on Lee Canyon Road. Reset odometer to 0.0 at the intersection of Lee Canyon Road (S.R. 156) and Deer Creek Road (S.R. 158).
	0.0	Continue straight ahead on the Lee Canyon Road (S.R. 156). *(GPS 72)*
1.2	1.2	Sawmill Trailhead and picnic area on left *(GPS 73)*
7.5	8.7	Leaving Spring Mountains National Recreation Area. *(GPS 74)*
0.3	9.0	The low hills ahead on the right (south) side of road display contorted and folded limestones of the Bird Spring Formation. The Sheep Range, with a maximum elevation of 9,912 feet at Hayford Peak, is at 11:00 to 2:00. Note the large alluvial fans at the base of the range. The Black Hills (small range just west of the Sheep Range) are nearly completely buried by the alluvial debris shed off the Sheep Range (refer to page 105 for a sketch of landforms typical in the arid Basin and Range province).
2.2	11.2	Note the playa (dry lake bed) at the south end of Three Lakes Valley in the distance at about 10:00. The rock layers of all mountain ranges visible ahead across the valley are tilted to the east at angles that range from 40 degrees to 70 degrees. These ranges were formed during basin-and-range extension (from about 17 to 6 million years ago in this area). Much of southern Nevada was moved westward relative to the stable Colorado Plateau during this major deformational event and the rocks of many mountain ranges were tilted to the east by movement along faults that were inclined to the west (see bottom figure on page 9 for a generalized section across the Basin and Range during this time).

Photo: Dr. Lloyd Glenn Ingles, California Academy of Sciences

Gambel's quail (*Callipepla gambelii*) have plump, stocky bodies, rounded wings, and prominent head plumes. The name Callipepla comes from the Greek kalli (beautiful) and peplos (robe). The body is gray above and buff below, with streaked sides. Males have a black face and throat and a white headband.

These birds spend most of their time on the ground, often near washes and springs. At night they roost in low branches of shrubs or trees. They are generally gregarious and form coveys of 20 or more in fall and winter.

Photo: Becky Purkey

Aerial view to the northwest over Corn Creek Flat showing the extensive white Pleistocene calcareous spring deposits. The deposits are currently being cut and washed away into the many branching and merging drainages that together make up Las Vegas Wash.

interval	cumulative	
3.0	14.2	Intersection with U.S. 95. Turn right to Las Vegas. *(GPS 75)*
0.3	14.5	Thick, white Pleistocene spring deposits of Corn Creek Flat are on the left at the foot of the Kyle Creek alluvial fan. These are the remains of extensive sheets of fine grained calcareous sediment deposited from springs along the ancient groundwater-surface interface in the valley bottom when the water table was higher than it is at present. The sediments were deposited during the two most recent glacial periods, one between 25,000 and 14,000 years ago and the other between 14,000 and 8,000 years ago. The wet meadows and marshlands associated with these springs were near the head of Las Vegas Creek, and probably merged with the Tule Springs-marsh-stream area to the southeast (described on Trip 5).
1.8	16.3	The Lee Canyon thrust fault lies within the prominent hill at about 3:00. The steeply northwest dipping (actually overturned) strata of the Bird Spring Formation have been thrust over gently northwest-tilted strata of the same formation. The thrust occurs at the boundary or contact between these differently oriented strata.

From here, return to Las Vegas on U.S. Highway 95.

End of Trip 2.

Color Photo Captions

PLATE 5

5a Cliffs of Mississippian limestone form the backdrop for bristlecone pines at the summit of **Mummy Mountain**. *Photo: Mark Vollmer*

5b Mississippian Monte Cristo Limestone massif of Echo Cliff, high above **Kyle Canyon** in the **Spring Mountains**. *Photo: Mark Vollmer*

5c **Bristlecone pine forest** atop Fletcher Peak in the Spring Mountains. Mummy Mountain is on the skyline in the background. *Photo: Mark Vollmer*

5d **Mule deer** (*Odocoileus hemionus*) prefer mixed habitats and forest edges and are mostly active mornings and evenings. They eat grasses and herbs when they are available but depend mainly on a variety of shrubs. *Photo: Roy W. Cazier*

5e **Greenleaf manzanita** (*Arctostaphylos patula*) provides an important browse for mule deer, and its berries are favored by a wide variety of birds and small mammals. *Photo: Joe Tingley*

5f,g,h **Quaking aspen** (*Populus tremuloides*). The quaking aspen is found in moist areas at the higher elevations of Nevada's mountains. Its name refers to its leaves which flutter with the slightest of air movement. *Photos 5f,h: Kris Pizarro; photo 5g: Jack Hursh*

PLATE 6

6a View, to the northeast, of the **Sheep Range** from State Route 158 in the Spring Mountains. The Sheep Range is on the right in the background, and the Black Hills protrude from the high valley to the left of the Sheep Range. The **Desert Range** is partly obscured by the large area of shadow that extends toward the Sheep Range from the left. *Photo: Mark Vollmer*

6b An outcrop along the road to the PABCO gypsum mine. **Gypsum** of the Muddy Creek Formation (flat-lying white bed) lies above **sandstone** and **siltstone** of the Thumb Member of the Horse Spring Formation (red beds that dip down to the right in the photo). The contact between the two formations is an **angular unconformity**. *Photo: Steve Castor*

6c Faulted blocks of gray limestone of the Kaibab Formation above red Aztec Sandstone. Oblique aerial view looking to the east above the **Redstone Picnic Area** on Northshore Road, Lake Mead National Recreation Area. *Photo: Becky Purkey*

6d Bristlecone pine atop **Mummy Mountain** in the **Spring Mountains**. The mountain ranges in the background are within the U.S. Air Force's Nevada Test and Training Range. *Photo: Mark Vollmer*

6e Branches of **bristlecone pine** displaying the smooth, whitish gray bark, the short, stiff needles that crowd against the branches, and the small cones (2½ to 3½ inches long) characteristic of the tree. *Photo: Kris Pizarro*

6f Closeup of a **bristlecone** trunk, showing the effects of centuries of weathering. *Photo: Mark Vollmer*

PLATE 5

PLATE 6

PLATE 7

PLATE 8

Color Photo Captions

PLATE 7

7a A **sunray** (*Enceliopsis spp.*) brightens a wash. **Lava Butte** dominates the horizon. *Photo: Jack Hursh*

7b **Bitter Spring Valley**. The washes and gullies are cut into the soft sedimentary rocks of the Thumb Member of the Horse Spring Formation that make up the valley floor. *Photo: Jack Hursh*

7c Black **basalt** boulders from an overlying basalt flow litter a slope cut on red Aztec Sandstone. *Photo: Becky Purkey*

7d **Mojave aster** (*Xylorhiza tortifolia*) thrives on rocky desert alluvial fans and washes. It has a lavender or pale violet flower and blooms in the spring and again in the fall, depending on rains. *Photos: Jack Hursh*

PLATE 8

8a A "**precariously balanced rock**" derived from Aztec Sandstone in Valley of Fire State Park. Similar to the hoodoos described in Trip 4, this balanced rock provides clues helpful in assessing the seismic risk of this area (see discussion on page 111). *Photo: Jack Hursh*

8b Brilliant red Aztec Sandstone exposed in the **Bowl of Fire**, Lake Mead National Recreation Area. *Photo: Jack Hursh*

8c, d, e, f, g, h Rock textures and patterns displayed by outcrops of Aztec Sandstone in Valley of Fire State Park. All six photos show **differential weathering** and the variegated coloring characteristic of the sandstone. **Rock varnish**, the shiny black coating that gives the rock surface the appearance of having been splashed with tar, is apparent in photos c, d, and f. Photos f and h display differential weathering controlled by bedding planes (the longer groves) and rock fractures (the less apparent groves at right angles to the bedding planes). In photo g, weathering along fairly closely spaced fractures is causing the rock face to crumble away as small blocks. *Photo 8c: Carol McKim; photo 8d: Patty Gray; photos 8e,f,g,h: David Gray*

8i Ancient Puebloan rock art in **Petroglyph Canyon**, Valley of Fire State Park. The dancing figures were chipped on a surface blackened by rock varnish. *Photo: Mark Vollmer*

interval cumulative

To find the starting point of this trip, from I-15 in north Las Vegas, take Exit 45 to Lake Mead Blvd. (State Route 147). Travel east on SR 147 for about 6.3 miles to the point where the road approaches the lower flanks of Frenchman Mountain and intersects Hollywood Blvd. This intersection could be termed "Pharmacy Corners;" Walgreens, CVS, and Rite Aid each occupy their respective corners, although the remaining fourth corner hosts a McDonalds. The shopping area centered about this intersection is the last you will encounter on this trip. If you need gas, water, or other staples, stop and stock up here.

0.0 Set your odometer to 0.0 and continue east on Lake Mead Boulevard. Frenchman Mountain now looms ahead and extends to the right; the lower hills to the left (north) of the road are part of Sunrise Mountain. *(GPS 76)*

1.0 1.0 The "Great Unconformity" interpretive site and trail is on the right. Before the site was extensively vandalized, there were informational signs located on the large stone monument at the parking area, and at points along the trail extending up the hill to the south. The monument south of the dry wash was erected at the foot of the hill, where the "Great Unconformity" between Cambrian sandstone and the Precambrian rock is well exposed. This feature was well illustrated and explained on the plaque attached to the monument, but it too has suffered from thoughtless vandalism. The rocks remain, however, and their relationships are described in the photos on page 56. *(GPS 77)*

To learn more, take the short trail (only about ¼ mile long, but rocky and steep) leading from the Great Unconformity interpretive site to an overlook point on the top of a hill to the south. The trail was built in the mid-1990s as part of an Eagle Scout project, and the now-destroyed signs were put in place as the result of a joint project of the Citizens for Active Management of Frenchman-Sunrise Area and the Las Vegas Field Office of the U.S. Bureau of Land Management. If you like, continue hiking to the east where you will see progressively younger Paleozoic rock layers stacked above the Cambrian sandstone.

When you have seen enough Paleozoic rocks, return to your car—but before you leave the parking area, look to the north of the road on the skyline and see if you can find the small natural "window" in the limestone that forms the ridgetop.

Turn right out of the parking area and continue east on Lake Mead Boulevard.

TRIP 3: FRENCHMAN MOUNTAIN AND VALLEY OF FIRE

This trip follows Lake Mead Boulevard (State Route 147) east from North Las Vegas through the low pass between Sunrise Mountain and Frenchman Mountain and into the Lake Mead National Recreation Area. Continuing through the Recreation Area, the route turns east to follow Northshore Road for about 40 miles through the broad desert washes and colorful rock outcrops that characterize lands along the north shore of Lake Mead. At St. Thomas Wash and Valley of Fire Road, there is the option to continue on to Overton and the Lost City Museum and return, or to turn west on Valley of Fire Road to travel through the Valley of Fire State Park and complete the trip back to Las Vegas via Interstate 15. The total mileage of the trip is about 120 miles. Gasoline is available at Overton, the halfway point on the trip.

The stretch of road beginning at Frenchman Mountain and extending along the north shore of Lake Mead to Valley of Fire passes through some of the most spectacular scenery and complex geology in the southwestern United States.

Just east of here, the Basin and Range province on the north and west meets the Colorado Plateau to the east. The rocks now exposed at this boundary tell the story of how the two provinces, a single geologic entity during long periods of deposition and erosion in a variety of Paleozoic and Mesozoic environments, were moved apart by extensional forces 17 to 10 million years ago during the late Cenozoic Era.

On this trip you will see effects of this extensional deformation; rock layers that have been tilted, folded, and broken by faults as the region was pulled apart and entire mountains that have been displaced tens of miles westward along some of these faults. One example is Frenchman Mountain, now occupying space directly east of Las Vegas. Rocks that make up Frenchman Mountain have been correlated with identical rocks situated east of Lake Mead, strong evidence that the mountain journeyed from there to its present location. Much volcanic activity also accompanied this period of crustal extension. Magma (molten rock) welled up and was expelled through vents and fissures in the Earth's crust to form lava flows and volcanoes. Features such as Lava Butte, east of Frenchman Mountain, and the lava flows seen on Black Mesa a few miles further to the east are evidence of this volcanism.

Early human activity within the area explored by this trip was concentrated along the valley of the Virgin River where American Indian village sites as old as 10,000 years have been found. Unfortunately, many of these sites are now submerged beneath the waters of Lake Mead. Although there were Spanish forays along the Colorado River as early as 1540, and a party of Spanish missionaries and explorers under Father Escalante may have crossed into Nevada in 1776, the first recorded European presence in the area was mountain man Jim Bridger who passed through in 1826. Not really impressed, in his journal he describes the "hilly barren appearance" and "terrible rocks" of the region. Two early U.S. Government expeditions traveled the Colorado River slightly south of the area of our trip. Lt. Joseph Ives with the Corps of Topographical Engineers, traveling upriver from the Gulf of California, reached the confluence with Las Vegas Wash in 1858. Major John Wesley Powell and his expedition party, traveling from Green River, Utah, reached the junction of the Colorado and Virgin Rivers in 1869 during their famous exploration of the upper Colorado River system.

Photo: Becky Purkey

The northern portion of Frenchman Mountain, view to the east from North Las Vegas.

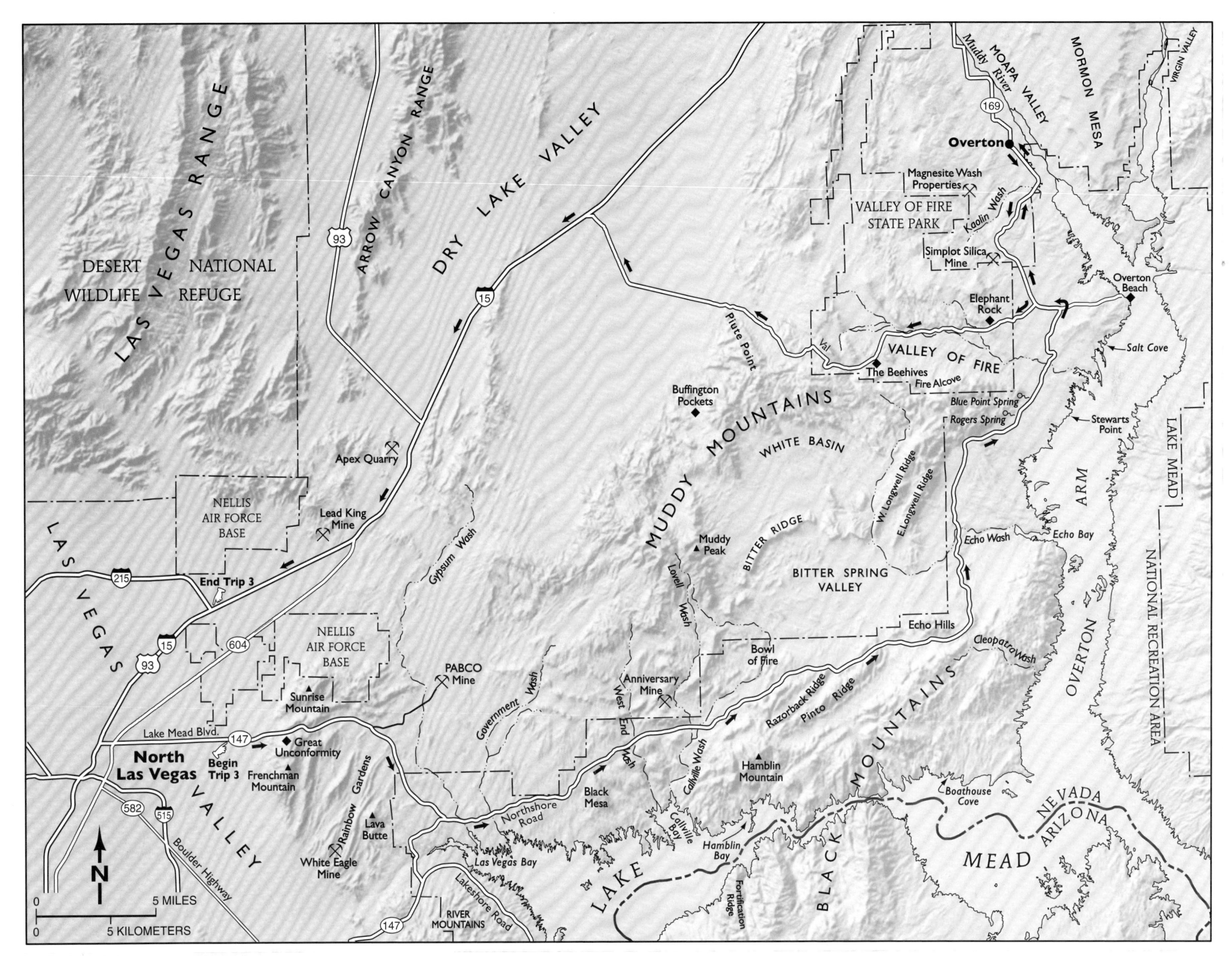

Route map, Trip 3.

The Great Unconformity

The rocks exposed in Frenchman Mountain range in age from 1.7 billion years (Precambrian) to as young as 5 million years (early Pliocene). The Precambrian rocks occur on the west side of the mountain at its base; the youngest rocks are on the east side of the mountain. Although there are some major gaps, the long time span represented by these rocks affords the most complete record available of the geologic history of this part of Nevada during the past 1.7 billion years.

At the western front of the mountain, rocks immediately to the right (south) of the road are Precambrian igneous and metamorphic rocks that are similar to the 1.7-billion-year-old rocks found in the inner gorge of the Grand Canyon. The Precambrian rocks exposed here at Frenchman Mountain are pink granite, pegmatite, and biotite (black mica) schist and banded gneiss composed mainly of feldspar, biotite, and quartz. Some samples of gneiss contain crystals of red garnet. Pegmatite is essentially granite with extremely large crystals which grew in pockets and veins by slow precipitation from gaseous fluids. Locally, the pegmatite contains large crystals of the mineral microcline (a variety of feldspar). Most of the microcline pieces scattered on the ground here are cleavage fragments that have been broken from even larger single crystals.

Resting on the Precambrian rocks are conspicuously layered and cross-bedded, brown rocks (refer to Trip 1 for an explanation of cross-beds) that are tilted to the east. The thin, cross-bedded rocks host numerous fossilized worm trails, burrows, and grazing patterns indicating that this was a marine beach environment during the time of deposition. These layered rocks are part of the Cambrian Tapeats Sandstone (about 520 million years old). Thus, there is a gap of more than one billion years in the geologic record between the Precambrian and Cambrian rocks at this locality. If rocks were deposited during this time interval, they must have been removed by erosion prior to the deposition of the Tapeats Sandstone, for we find no trace of them in southern Nevada. This situation, a contact between much younger rocks deposited directly on older rocks where the intervening rocks either were never present or were removed by erosion, documents a gap in the geologic record and is called an unconformity.

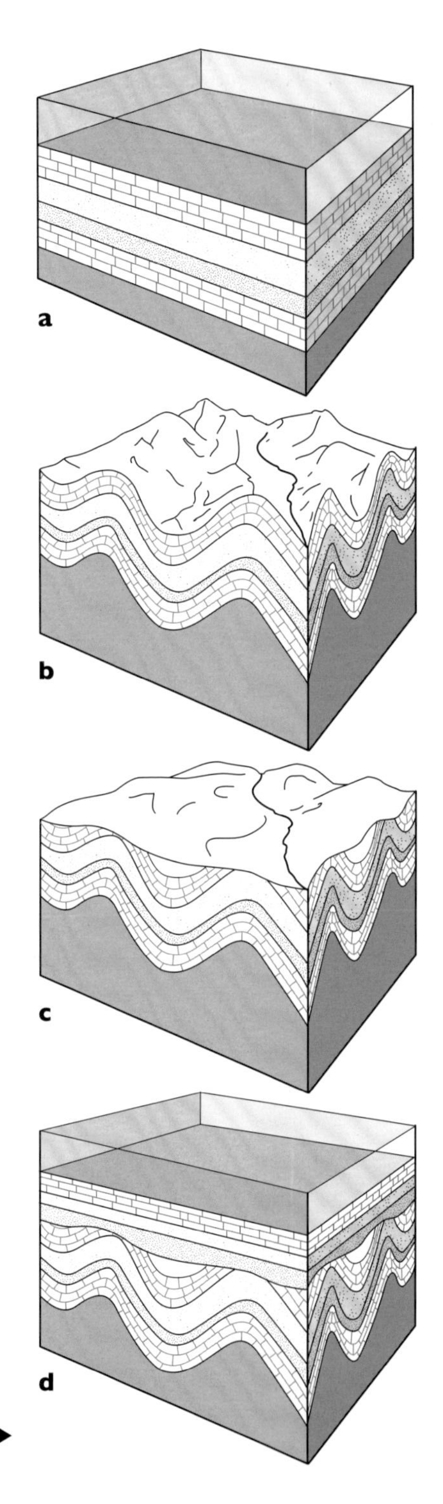

Development of an angular unconformity a) deposition of sediments, which are consolidated into sedimentary rocks; b) folding of rocks, erosion, c) continued erosion, producing a "gap" in the geologic record; d) deposition of younger layered rocks on the eroded surface of the folded older rocks. ▶

Photo: Becky Purkey

Aerial view, to the southwest, of Sunrise Mountain (foreground), and Frenchman Mountain (middle background). Lake Mead Blvd. runs east and west between the two mountains. Las Vegas Valley is in the background.

Photo: Becky Purkey

A close-up of the contact between Precambrian metamorphic and igneous rocks and overlying, tilted Cambrian Tapeats Sandstone at the "Great Unconformity."

Photo: Becky Purkey

The beginning of the Great Unconformity interpretative trail. Precambrian granite (rubbly rock) extends from the marker to the right; Cambrian Tapeats Sandstone (layered rock) extends from the marker to the left. The trail angles up to the right across the hillside.

interval cumulative

0.5 1.5 Most of the rocks to the left, forming the ridge top, are Permian Kaibab Formation. The rocks have been cut by several faults and are generally dipping to east. There are chert lenses and nodules along bedding (the individual rock layers), giving the rock its brown, ribboned appearance. *(GPS 78)*

0.5 2.0 To the left of the road, notice that the rock layers are not tilted east as you saw at the "Great Unconformity" and up to about mileage 1.5, but have been bent around to a more southerly tilt. This bending, sort of hard to see while driving by but very apparent from the air, is probably a consequence of movement along the Las Vegas Valley shear zone, a major right-lateral strike-slip fault that lies north of Sunrise Mountain. Strike-slip faults form where crustal blocks slide horizontally past one another. The "lateral" part of the description tells which way the faulted blocks moved. An easy way to keep the movement straight is to visualize yourself standing on a block of ground and looking across the fault at the other side. If you need to look to the right to find the offset piece of the block you are standing on, it is a right-lateral fault, if you need to look left to find the offset piece, it is a left-lateral fault.

0.3 2.3 Look on the skyline at about 11:00 for what appears to be a dark cave with an arched roof. As you drive east, keep watching and you will see daylight through what is actually a natural arch in the Permian limestone that makes up the ridge (also keep your eyes on the road here as the road curves). *(GPS 79)*

1.2 3.5 At 3:00 there are numerous small caves formed in the limestone outcrop.

0.3 3.8 Bright-red Jurassic Aztec Sandstone (about 180 million years old) crops out at the base of the ridge to the south of the road at 3:00. This is the same rock layer that is exposed in the Red Rock Canyon area west of Las Vegas (see Trip 1) and in the Valley of Fire (ahead on this trip). The Aztec Sandstone is an extensive unit of ancient lithified sand dunes found throughout the southwestern United States that formed in an environment much like the modern Sahara Desert. The ridge is capped by the Rainbow Gardens Member of the Horse Spring Formation. *(GPS 80)*

Closer to the road, the softer-appearing red hills are composed of the Triassic Kayenta Formation, the unit directly beneath the Aztec Sandstone. Here the Kayenta contains some gypsum which weathers out to form the small clear crystal shards you see glistening in the sunlight.

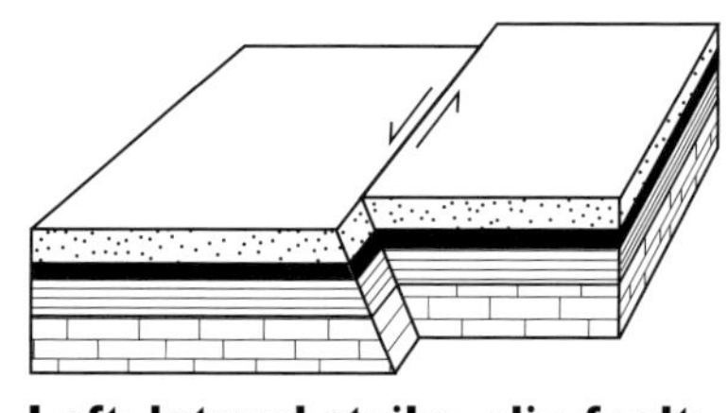

Left-lateral strike-slip fault

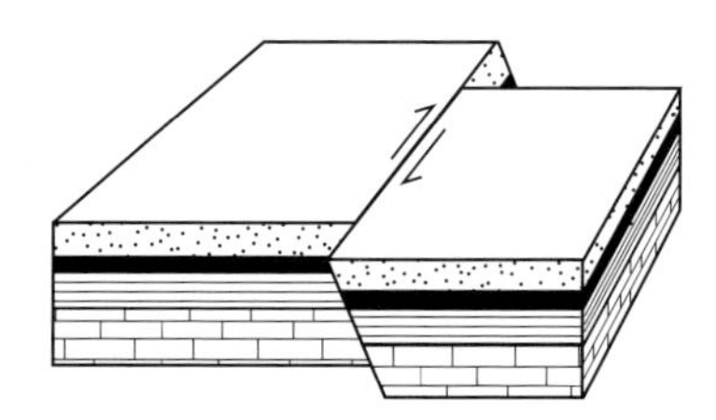

Right-lateral strike-slip fault

Left-lateral strike-slip fault. Each block appears to have been shifted left relative to the other. Right-lateral strike-slip fault. Each block appears to have been shifted right relative to the other.

GEOLOGIC PERIOD	FORMATION NAME	CORRELATING UNIT ON GEOLOGIC MAP (See page 13)
Quaternary	**Alluvial deposits** (sand, gravel, including alluvial fan and terrace gravel deposits)	Quaternary Alluvium
— 1.8 Ma Tertiary	**Muddy Creek Formation** (sandstone, siltstone, shale, and conglomerate, including gypsum layers)	Tertiary Muddy Creek Formation
~ 10-8.5 Ma	**Red sandstone unit** (sandstone, siltstone, and minor tuff (volcanic ash))	not shown
— 13 Ma — 17 Ma — 65 Ma	**Horse Spring Formation** **Lovell Wash Member** (interbedded limestone, sandstone, siltstone, and tuff) **Bitter Ridge Limestone Member** (mostly limestone with local sandstone, sandy limestone, gypsum, and borate) **Thumb Member** (mostly sandstone with beds of conglomerate and gypsum) **Rainbow Gardens Member** (limestone and conglomerate at the base, overlain by limestone, conglomerate, dolomite, local sandstone, and magnesite (in the Overton area))	Tertiary Horse Spring Formation
Cretaceous	**Baseline Sandstone, including Willow Tank Formation** (white and gray, well-bedded sandstone with lenses of conglomerate)	Cretaceous Baseline Sandstone and Willow Tank Formation
~~	*unconformity*	
Jurassic	**Aztec Sandstone** (red, tan, or yellow cross-bedded sandstone)	Jurassic Aztec Sandstone
— 206 Ma Triassic	**Kayenta Formation** (brick red sandstone and siltstone) **Chinle Formation** (dark red to purple siltstone, mudstone, minor sandstone, and conglomerate) **Moenkopi Formation, including Virgin Limestone Member** (reddish siltstone, shale, and sandstone, with a prominent limestone member - the Virgin Limestone)	Triassic rocks, undivided
~ 248 Ma	*unconformity*	
Permian	**Kaibab and Toroweap Formations, undivided** (gray limestone with intervals of white gypsum) **Permian red beds** (red sandstone and siltstone)	Permian rocks, undivided
— 290 Ma Pennsylvanian	**Bird Spring Formation** (gray limestone and dolomite)	
— 323 Ma Cambrian through Pennsylvanian	(limestone, dolomite, calcareous shale, and some quartzite)	Cambrian through Pennsylvanian rocks, undivided (excluding Bird Spring Fm. and Tapeats Sandstone)
Cambrian	**Tapeats Sandstone** (brown, cross-bedded sandstone)	
~~	*unconformity*, Silurian rocks missing	
Proterozoic — 2,500 Ma	(pink granite, pegmatite, biotite, schist, and banded gneiss)	Precambrian granite and metamorphic rocks

Numbers followed by "Ma" indicate millions of years ago.

Generalized stratigraphic column for the Frenchman Mountain-Valley of Fire area.

THE JOURNEY OF FRENCHMAN MOUNTAIN

The Paleozoic rocks (540 to 248 million years old) on Frenchman Mountain are different from Paleozoic rocks in other mountains around Las Vegas, but are very much like rocks near the Grand Canyon, more than 60 miles to the east. This fact, plus the presence of some unique landslide deposits found on the east side of Frenchman Mountain, led Yale University (and U.S. Geological Survey) geologist Chester Longwell to propose, in the early 1970s, that Frenchman Mountain was once located somewhere in the vicinity of Gold Butte, east of present-day Lake Mead, and was moved along faults to its present position during late Cenozoic time (about 17 to 10 million years ago).

Longwell thought that the Las Vegas Valley shear zone, a major right-lateral strike-slip fault that lies along the northern margin of Las Vegas Valley, was responsible for this transport. A decade later, R.E. Anderson and Robert Bohannon of the U.S. Geological Survey also proposed the Gold Butte area as a likely launching pad for Frenchman Mountain, but suggested that the 40-some miles of transport was along the left-lateral Lake Mead fault system (see page 8 for definitions of the types of faults).

A key player in the evidence trail of this story is the sequence of sedimentary rocks known as the Horse Spring Formation. These rocks were probably deposited in shallow, periodic lakes formed in the large basins produced during the early stages of Cenozoic extension (probably starting about 17 million years ago).

The landslide deposits that provided Longwell with one of the clues to the Frenchman Mountain puzzle are preserved as lenses of breccia within the Thumb Member of the Miocene Horse Spring Formation. These breccias contain large blocks (up to building-size) and fragments of Precambrian gneiss and rapakivi granite (a dark, coarse-grained granite with very large feldspar crystals). However, no rocks exposed near Frenchman Mountain today could have been a source for these breccias. The nearest known outcrops of rapakivi-type granite are near Gold Butte in the South Virgin Mountains. Since this distance is too far for a landslide containing rocks of this size to travel, geologists think that the breccias must have been deposited close to Gold Butte. Then the entire block (including Frenchman Mountain and the telltale breccia) was sent on its way along a fault-guided path, eventually coming to rest to the east of present-day Las Vegas.

Adding to the story, conglomerate in the Bitter Ridge Limestone Member of the Horse Spring Formation, the unit that overlies the Thumb Member, contains fragments of Paleozoic limestone thought to have washed in from a source area to the south of the shear zone. Since now there are no rocks like this south of the shear zone, more evidence for extension piles up. Frenchman Mountain, with its Paleozoic rocks, was in position about 13 million years ago to have shed limestone into the Bitter Ridge Basin as it passed by on its way west.

Currently, most geologists agree that Frenchman Mountain is "out-of-place," but controversy exists among these same geologists regarding the exact original position of the mountain, the faults that caused movement of the block, and the period of time in which this movement occurred.

Photo: Becky Purkey

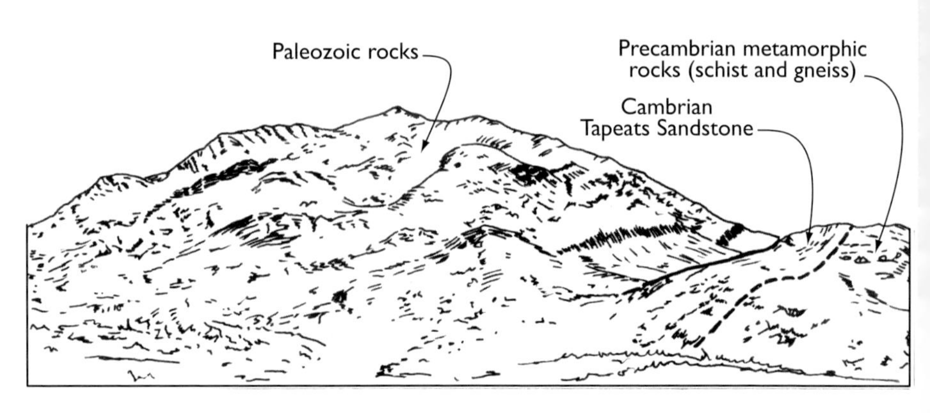

Frenchman Mountain, the view in the photo is to the southeast from Lake Mead Blvd. The sketch shows the general stratigraphic units exposed in this part of Frenchman Mountain.

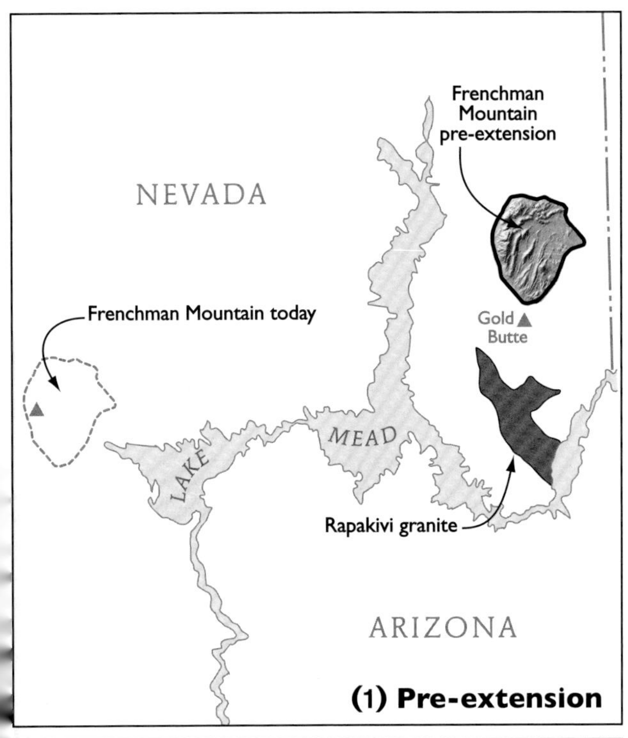

(1) Pre-extension

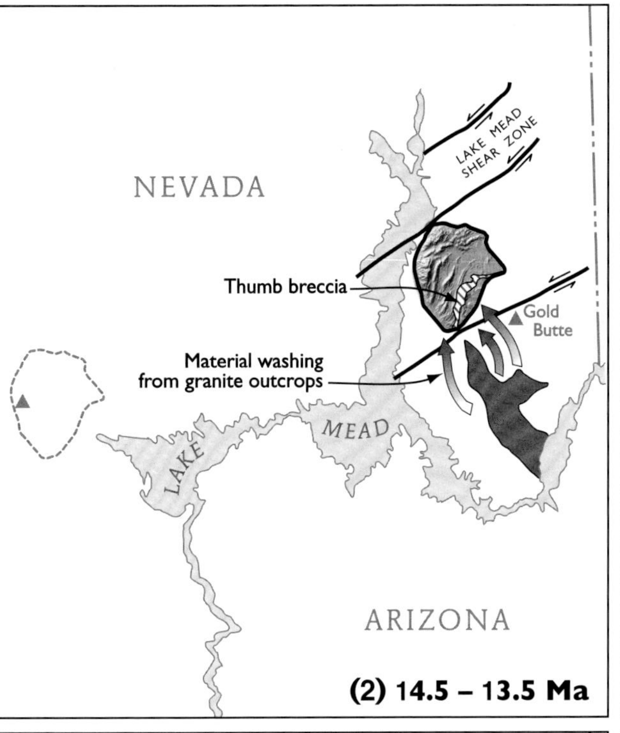

(2) 14.5 – 13.5 Ma

(3) 13.5 – 12.5 Ma

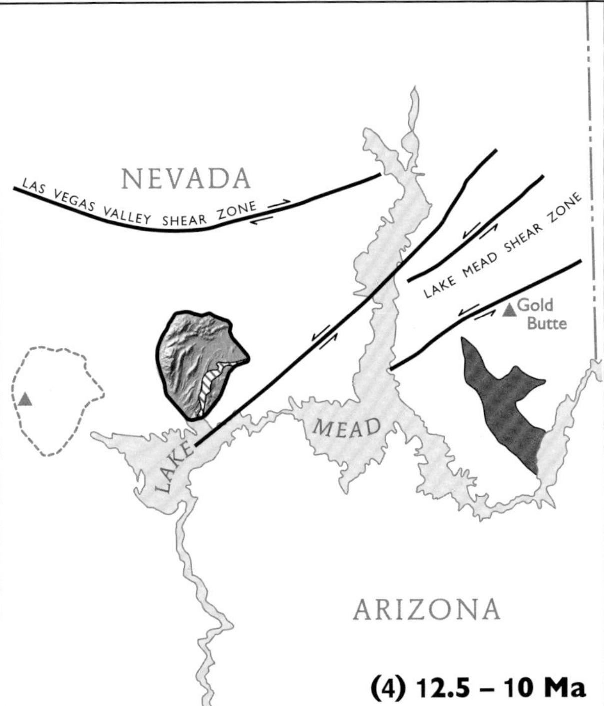

(4) 12.5 – 10 Ma

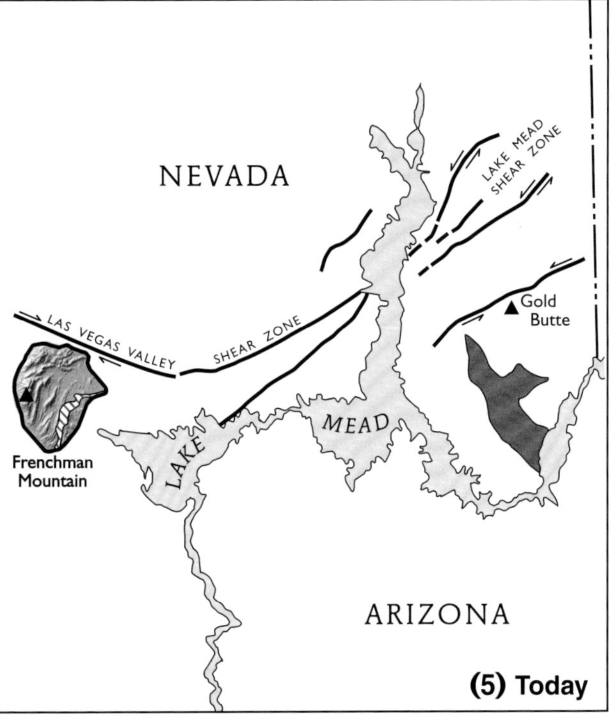

(5) Today

Travels of Frenchman Mountain:

1. Frenchman Mountain, ready to travel, in its pre-extension position somewhere north of present-day Gold Butte. Precambrian rapakivi granite is also in place to the south.

2. Extension is underway, tilting along the developing Lake Mead shear zone causes rapakivi blocks and boulders to be washed into the basin collecting the sediments that we now call the Thumb breccia, a part of the Thumb Member of the Horse Spring Formation.

3. Now we see the effects of some serious movement along the Lake Mead shear zone. The tectonic block we now know as Frenchman Mountain is about halfway to its present position. Along the way, however, it is weathering and shedding Paleozoic limestone debris north into what is now Bitter Spring Valley.

4. Our traveler is now close to its future home. The Las Vegas Valley shear zone is now in the picture, and some or all of the continuing journey of Frenchman Mountain may be due to movement along this shear zone.

5. This is what we see today, Frenchman Mountain settled into its present position—but for how long? Extension is still in progress and Frenchman Mountain could just be here for a geologic "overnight."

interval	cumulative	
1.2	5.0	The road to PABCO Gypsum's gypsum mine and plant is on the left. The operation is about 3 miles on paved road from this intersection, but is not open to the public. In the basin to the east, the road crosses through lots of colorful outcrops of Miocene-Pliocene Muddy Creek Formation, and in places crystals of selenite gypsum litter the land surface. If you decide to tour this area, stay away from the private property and active mining areas, watch for trucks, and don't drive off of the pavement. *(GPS 81)*
0.1	5.1	The low ridge on the right-hand side of the road is composed of conglomerates of the Horse Spring Formation. This formation forms extensive outcrops in the large basin cut by Gypsum Wash to the left, and we will see much more of it along our route ahead.
0.5	5.6	The small hill to the right at about 2:00 (next to the road) is capped by Thumb breccia (part of the Thumb Member of the Horse Spring Formation). There is a nice wide turnout here, so if you want to get a close look at this special rock unit, this is a good place to stop. An easy walk will take you to the flank of the hill where boulders rolling from the top contain fragments of rapakivi granite and gneiss. *(GPS 82)*
0.2	5.8	The prominent dark peak at 2:30 is Lava Butte. Lava Butte is composed of a volcanic rock called dacite, which has been dated at about 13 million years; so these rocks are very much younger than the Paleozoic sedimentary rocks of Frenchman Mountain. On a fresh, broken surface, the dacite of Lava Butte is quite attractive. It is light gray to lavender and contains visible crystals of mica, feldspar, and hornblende. The conspicuous dark coloration is rock varnish, a surficial coating of clay and manganese and iron oxides (see page 85 for a discussion of rock varnish). *(GPS 83)*
		Lava Butte is not a volcano. Recent mapping has led geologists to interpret it as a small volcanic plug (a laccolith, see sketch on page to the right) that came up through the Horse Spring Formation, forcing its way laterally along the formation for a short distance at one or more horizons. It now stands up as a prominent peak above its surroundings because it is composed of rocks that are more resistant to erosion than the soft sedimentary rocks that it forcefully penetrated.
		The low ridge between the highway and Lava Butte has red Aztec Sandstone cropping out at its base and is capped by the Rainbow Gardens Member of the Horse Spring Formation (the oldest member of the Horse Spring, it was deposited at least 17 million years ago). This same relationship is seen in the extension of the ridge to the left of the highway at about 11:00. Follow the trend of the ridge to the left, beyond the transmission tower and you will see a large hill capped with dark volcanic rock. Most of the basin to the left is underlain by the Thumb Member of the Horse Spring Formation (about 16 to 12.5 million years old), but the white sediments on the far side of the basin are gypsum-bearing beds of the Muddy Creek Formation (10 to 5 million years old). The PABCO plant is at 9:30.

PABCO Gypsum

Gypsum is a mineral (hydrated calcium sulfate) formed by evaporative deposition from a body of water, a process in which dissolved salts become so concentrated that small crystals precipitate out of solution and settle to the bottom of the sea or lake. The PABCO gypsum deposit, unlike many deposits in Nevada that formed in a marine environment (such as the Blue Diamond gypsum deposit seen on Trip 1), was probably precipitated in a desert playa lake.

The gypsum is mined from a deposit in the 10- to 5.0-million-year-old Muddy Creek Formation. Some fine-grained sugary gypsum is present, but most commonly the gypsum occurs as white to clear selenite crystals interbedded with lenses of clay. The gypsum ore is mined in an open pit and hauled to a conveyor system that moves it to a washing plant where a 92 to 95 percent gypsum product is obtained. This material is calcined (roasted to drive off some of the water in the crystal structure of gypsum, changing it to plaster of Paris) and either shipped or used here to make wallboard.

Limited mining was done in this area in the 1940s, and perhaps earlier, and the deposit has been continuously mined by PABCO since 1959. In 2000 this mine produced about 937,000 tons of gypsum and was the largest of Nevada's gypsum mines.

If you are a James Bond follower, you might recognize the scenery in the vicinity of the PABCO plant. This mill was used in the 1971 James Bond film "Diamonds are Forever." Some of the outdoor action was filmed in the area between our route and the road to the gypsum plant—Lava Butte is even recognizable in the background in one of the chase scenes.

Photo: Steve Castor

PABCO gypsum processing plant.

interval cumulative

0.5 6.3 This is a good place to see the unconformity between the Aztec Sandstone (about 180 million years old) and the Miocene Rainbow Gardens Member of the Horse Spring Formation (at least 17 million years old). This is a normal depositional contact between the two rocks, but it is called an unconformity because all rocks between the Jurassic and Miocene time are missing—they were eroded away before deposition of the Rainbow Gardens Member. For the next quarter of a mile, we will be traveling up section through the Miocene rocks (from older rocks up through progressively younger rocks); the sequence is basal conglomerate, a sandstone-limestone unit, and then a resistant limestone unit. *(GPS 84)*

0.5 6.8 More Thumb breccia to be seen here, both the prominent hill just south of the road to the right, and the low hill holding up the two power transmission towers to the left of the highway are capped by this breccia. If you missed it at mileage 5.6, or want to stretch your legs some more, you can hike to the top of the butte here and see the Precambrian rock fragments in the breccia. Check for crystalline rock with large feldspar crystals—they have cleavage surfaces that reflect light. *(GPS 85)*

The wide gravel road to the right leads to Rainbow Gardens, an area of colorful rock formations that lies between the highway and the eastern base of Frenchman Mountain. The Rainbow Gardens Member of the Horse Spring Formation takes its name from this area and, in turn, provides much of the color. Exposures of banded red, white, and tan-colored rocks in the closest hogback ridge beyond the wash to the west are typical of this formation.

Rainbow Gardens is within the Bureau of Land Management's Sunrise Mountain Management Area, and a BLM informational sign about 200 feet west of the highway presents a good overview of the local geology. Branching roads lead to Lava Butte and, after about 6 miles, the road passes an abandoned gypsum mine (the White Eagle Mine) and then drops back to the west into Las Vegas Valley. This road, although passable to high-clearance vehicles, should not be attempted by most passenger cars. In flash flood season, be especially cautious when crossing the large wash about 0.3 miles from the highway.

Lava Butte. View to the south from Northshore Road.

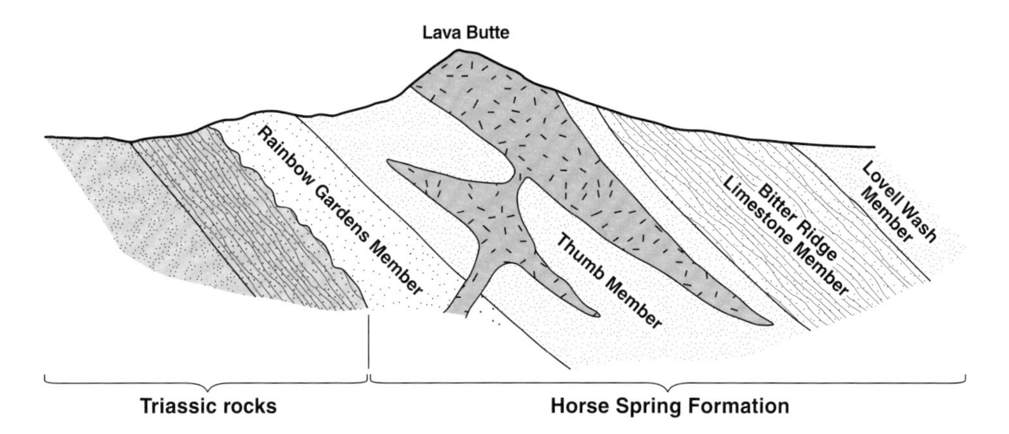

Generalized cross section of Lava Butte. Lava Butte is a laccolith, a type of intrusion formed along bedding layers in the host formation, resulting in the flat lower intrusive contact shown in the sketch. (modified from Castor and others, 2000)

Photo: Joe Tingley

Closeup of a block of Thumb breccia that has broken from the top of the hill in the photo to the right. The striped rock fragment below the lens cap is Precambrian gneiss; the smaller, light-colored fragments scattered in the background right of the lens cap are feldspar from rapakivi granite.

Photo: Steve Castor

Small hill composed of the Thumb Member of the Horse Spring Formation. The hill is capped with Thumb breccia, a unit within the formation containing blocks and fragments of Precambrian gneiss and rapakivi granite.

Photo: Jim Faulds

Aerial view, to the north, of Rainbow Gardens. Frenchman Mountain is in the upper left, Lava Butte is on the right. The area of light-colored rocks in the center of the photo is Rainbow Gardens. The white patch where the curving road from the lower left enters the range is the White Eagle gypsum mine.

interval	cumulative	
0.8	7.6	Large lenses of breccia containing blocks of Precambrian gneiss and rapakivi granite occurring within of our now-familiar Thumb Member of the Horse Spring Formation can be seen in the small gray hills to the right and left sides of the road.
0.2	7.8	Note the unnamed wash at right. This route crosses many washes (also called gullies or arroyos) which are tributary streams to the Colorado River. These washes are usually dry, but they were formed by rushing water during cloudbursts and resulting flash floods, and some may have been permanent streams thousands of years ago during wetter periods. They usually are cut through softer sediments and have distinctive gravel-covered flat bottoms and steep walls. The gravel moves along the bottom of a wash during floods, undercutting the softer sediments along the banks and causing them to collapse into the moving stream. Thus a wash maintains a flat bottom and grows wider while the banks remain steep and relatively parallel.
0.3	8.1	The large sign on the right marks the entrance to Lake Mead National Recreation Area. Lava Butte is at 2:30. The low, rugged ridge that extends northeast toward you from the butte is a narrow extension of the Lava Butte intrusive rock. Notice that the extension weathers much the same as the main butte, shedding a skirt of dark debris. *(GPS 86)*
0.2	8.3	The white rocks on both sides of the road are part of the 13 million-year-old Lovell Wash Member of the Horse Spring Formation. This is the uppermost or youngest member of the Horse Spring Formation. The white color of these rocks is due to the large amount of volcanic ash they contain. The source of the ash, which must have resulted from major explosive volcanic eruptions, has not been identified.
0.3	8.6	Black Mesa can be seen at 10:00 in the distance. At 11:00, beyond the low hills in the foreground, you can see the top of Fortification Hill located across Lake Mead in Arizona.
0.1	8.7	Fee collection booth, Lake Mead National Recreation Area. *(GPS 87)*
0.2	8.9	The road again parallels the large unnamed wash on the right.

Photo: Becky Purkey

Unnamed wash during a flash flood in March 1992.

Photo: Becky Purkey

Black Mesa, ahead to the east along Northshore Road.

interval	cumulative	
0.4	9.3	Panoramic view, left to right: *(GPS 88)*

The Muddy Mountains, at 10:00 and slightly obscured by the road cut on the left, contain Paleozoic rocks cut by late Mesozoic thrust faults (about 100 to 90 million years ago) and are geologically similar to the Spring Mountains west of Las Vegas. This geologic similarity has led some geologists to propose that the two ranges were once part of a single geologic block before Cenozoic extension (starting about 17 million years ago) pulled them apart.

Black Mesa, the long, low hill with the flat top at 11:30, is composed of basaltic andesite that was erupted from fissures and cinder cones between 10.6 and 8.5 million years ago. These lava flows are restricted to a relatively small area east of Frenchman Mountain and north of Lake Mead. Hamblin Mountain is visible over the top of Black Mesa.

Across the lake in the northern Black Mountains of Arizona, Fortification Hill is the dark mesa with the sloping top at about 1:30. Wilson Ridge is on the skyline behind Fortification Hill, and the lower ridge to the left is Fortification Ridge. The Black Mountains consist of Precambrian metamorphic and igneous rocks that were intruded by granites of the Wilson Ridge pluton 14 to 12 million years ago.

The River Mountains, at 2:30 to 3:00, are described in more detail in Trip 4. This range is composed of the remnants of a major volcanic complex (stratovolcano and surrounding domes) that was active between 14 and 12 million years ago. (Refer to page 92 for diagrams of volcanoes.)

interval	cumulative	
0.7	10.0	Crossing a major unnamed wash.
0.4	10.4	Lava Butte is on the right at 3:00.
0.2	10.6	Intersection with Northshore Road. Turn left (east) and proceed toward Valley of Fire and Overton. *(GPS 89)*
0.4	11.0	Notice the dark volcanic rock (vesicular basalt) exposed in the road cuts along this section of road.
0.4	11.4	The paved turnout on the right provides a good view of Lake Mead with Fortification Hill in the background at about 1:30. Las Vegas Bay Marina is visible just beyond the rolling hills at 3:00 (look for trees!), and the River Mountains are in the background beyond the marina. This is a good place to view the rugged volcanic core of the River Mountains. The mountains are described on Trip 4, but you have a much better view from here. *(GPS 90)*

Photo: Carol McKim

White-tailed antelope ground squirrel *(Ammospermophilus leucurus)* is gray with a narrow white stripe along each side. The underside of its tail is white and is carried over its back. One of the few small mammals to be active in the heat of the day, this animal is superbly adapted for desert life:

- It is omnivorous and prefers water-filled plants and insects.
- It tolerates a nine-degree fluctuation (100–109° F) in body temperature as it moves in and out of the sun.
- It can salivate heavily and spread moisture over its neck and cheek for evaporative cooling effect.
- It can lose heat by conduction when it retires to the shade or its burrow and flattens itself against the ground.
- It will position itself parallel to the Sun's rays and shade its body with its tail.

interval	cumulative	
0.3	11.7	The dirt road to the left leads to upper Gypsum Wash. *(GPS 91)*
0.2	11.9	The road to the right (south) provides lake access (by foot only, no vehicles).
0.1	12.0	Cross Gypsum Wash. On the left (north) side of road is an unconformity between tilted red sandstone and siltstone of the Muddy Creek Formation (about 5 to 10 million years old) and flat-lying tan Quaternary sediments (1 to 2 million years old). This geologic contact provides evidence of the history of uplift, erosion, and deposition in the area in the last 10 million years. *(GPS 92)*
0.4	12.4	Paved turnout. There is a good view of Las Vegas Bay Marina with the River Mountains in the background at 3:00. *(GPS 93)*
0.7	13.1	Government Wash turnoff is on the right. Government Wash is crossed just beyond the turnoff. Proceed straight ahead (east) on Northshore Road. At 2:30, on the far side of Lake Mead, the Black Mountains are in the background and Fortification Hill is in the foreground. Fortification Hill is a remnant of extensive Miocene lava flows (about 6 million years old) that poured out over the area and are the youngest volcanic rocks in the Lake Mead region. Several volcanic centers and more than 100 individual lava flows have been identified on Fortification Hill. *(GPS 94)*
0.9	14.0	Photo turnout on the right. For a good view of a spectacular unnamed wash with Lake Mead in the background, pull into this large, paved loop turnout and parking area. Looking into the wash from the turnout, you can see horizontal beds of the Muddy Creek Formation exposed in the banks of the wash and as capping on large, haystack-shaped pillars in the wash. If you want to walk into the wash (don't do this in flash flood season), you can see the thick mats of gypsum-bearing rock that occur in this formation (look for white crystalline material at the base of the pillar of rock in the wash just south of the highway crossing). *(GPS 95)*
0.2	14.2	Road crosses unnamed wash.
0.5	14.7	The dark brown, rugged low hills in the middle distance at 9:00 (they pass behind the road cut on the left at mile 14.8, but come back into view at 14.9) are composed of late Paleozoic rocks (including the Permian Kaibab Formation) surrounded by younger Cenozoic rocks of the Horse Spring and Muddy Creek Formations. These rocks mark the position of the Las Vegas Valley shear zone—the major, right-lateral fault zone that bounds Las Vegas Valley on the north.

Photo: Becky Purkey

Unconformity between tilted beds of the Miocene Muddy Creek Formation (darker beds partially hidden by big blocks in the lower part of the bank) and Quaternary sediments (flat-lying, lighter-colored sediments at the top of the bank). Photo of Gypsum Wash at Northshore Road.

Photo: Joe Tingley

A haystack-shaped rock pillar capped with light-colored, gypsum-rich sediments of the Muddy Creek Formation in an unnamed wash east of Government Wash. The same light-colored sediments are exposed in the lower part of the bank of the wash in the background.

interval	cumulative	
0.6	15.3	Paved turnout on the left, Black Mesa lies directly ahead. The mesa is capped by a series of basaltic andesite lava flows known to geologists as the volcanic rocks of Callville Mesa. These flows erupted between 10 and 8 million years ago from cinder cones located in the low hills to the east between Black Mesa and West End Wash. *(GPS 96)*
		Features such as Black Mesa form where rocks that are resistant to erosion (basaltic andesite in this case) overlie weaker rocks (siltstone in this case). The weak rocks on the flanks of the mesa erode readily, but the resistant caprock serves as a barrier that slows the erosional process. The softer, underlying rocks are light-colored, but the entire mesa—sides and all—appear dark because the slopes are mantled by a layer of dark rock talus cascading down from the weathering cap rock.
0.3	15.6	Paved turnout on the right. There is a good view from here of Black Island in Lake Mead. Sentinel Island is beyond Black Island. From here, these islands line up to point to the entrance to Black Canyon, the former channel of the Colorado River now blocked by Hoover Dam. *(GPS 97)*
0.7	16.3	The turnoff to Boxcar Cove and Crawdad Cove is on the right. Both roads are usually negotiable by ordinary passenger vehicle. Be careful though because these roads are susceptible to flash flood damage. *(GPS 98)*
0.7	17.0	Road cut between two low hills. These hills are composed of landslide and debris flow deposits of the Thumb Member of the Horse Spring Formation. There is no safe place to stop here to look at this rock, but there are some very large granite and gneiss blocks exposed in the road cuts that you can probably see without stopping.
0.2	17.2	The multicolored rocks on the right side (south) of the road are part of the younger members of the Horse Spring Formation. The white rocks contain significant amounts of volcanic ash; the yellow rocks are limestone, and the black rocks capping the yellow layer are basalt flows.
0.2	17.4	Large, paved turnout here, rocks of the Horse Spring Formation crop out on the right and in the cut ahead. The hills across the road to the left are capped with breccia of the Thumb Member of the Horse Spring Formation. The breccia here is especially rich in rapakivi granite with large feldspar crystals, some up to ½ inch across. *(GPS 99)*
		Black Mesa looms on the skyline to the right.

Photo: Dr. Lloyd Glenn Ingles, California Academy of Sciences

A member of the cuckoo family, the **greater roadrunner** (*Geococcyx californianus*) is zygodactylous (two toes pointing forward and two backward). It has heavily streaked plumage, a bristly crest, short, rounded wings, an upward-tilted tail, and long legs. It rarely flies and has been clocked running at speeds up to 17 miles per hour. It is said to have gotten its name in the Old West from running ahead of horse-drawn vehicles. Its diet of lizards, insects and spiders, small rodents, birds, and snakes is supplemented with occasional fruits and seeds.

interval	cumulative	
0.9	18.3	More colorful rocks of the Horse Spring Formation are exposed in the wash on both sides of the road.
0.2	18.5	The Callville Bay turnoff is on the right. Callville Bay is a popular launching point for boats onto Lake Mead. Services include a small store, telephone, and campground. *(GPS 100)*
1.1	19.6	Entering a geologically complex area near the intersection of the region's two major strike-slip fault systems: the right-lateral Las Vegas Valley shear zone and the left-lateral Lake Mead fault system. In this area of intersection of the major faults, the rocks have been sliced, tilted, thrust over each other, and folded.
0.5	20.1	Paved turnout on the right. The flat-topped hill to the right (with road scars) is referred to by some as "Callville Mesa," although this is not an official name (see Trip 1, page 21, for a discussion on official names). The mesa is capped by volcanic rock which also caps the hill to the left of the road. These rocks are late Miocene age (10 to 8 million years old) and are called the volcanic rocks of Callville Mesa. *(GPS 101)*

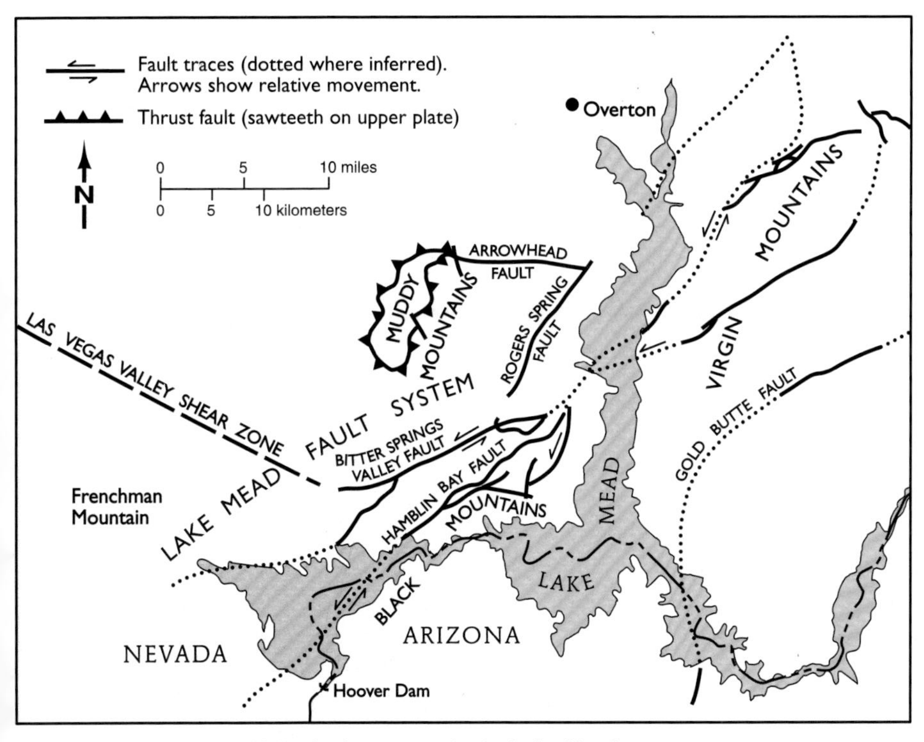

Major fault systems in the Lake Mead area.

CALLVILLE AND COMMERCE ON THE COLORADO

The idea of travel by river steamer up the Colorado took shape soon after the 1849 rush to the California gold fields sparked a need for quicker, cheaper transcontinental transportation. The first steamer, the Uncle Sam, set out upriver from Yuma, Arizona in 1852, and a brisk river trade between mines and settlements along the lower stretch of the river soon developed. In 1858, a U.S. Corps of Engineers expedition successfully steamed upriver to the point where Las Vegas Wash merged with the Colorado (misled by the size of the wash, the expedition leader believed he had reached the Virgin River even though the wash carried only a trickle of water). The success of the adventure attracted the attention of Mormon leader Brigham Young in Salt Lake City. At this time, freighting over the Mormon road from Los Angeles to Salt Lake City was a $3 million a year business, and the Mormon business community felt that substituting shipment by river steamer for part of the route could cut the freighting rate by a third. The Deseret Mercantile Association and Brigham Young sent Anson Call to find a suitable landing and freight transfer spot, and he established the Mormon settlement of Callville in 1864 on the west bank of the Colorado River about six miles above Las Vegas Wash. For a short time, steam-powered boats battled the rapids through Black Canyon, where Hoover Dam now stands, to Callville to bring supplies and people up river to connect with the overland route to the Virgin Valley and Utah. The Salt Lake trade never really developed, however, and when the transcontinental railroad was completed through Utah in 1869, the port was abandoned. The historical remains of Callville now lie beneath several hundred feet of water in Lake Mead.

Stone walls of an old warehouse building at the site of Callville on the Colorado River. Photo taken in about 1930. *Photo: Nevada Historical Society*

interval	cumulative	
0.2	20.3	Note the dark basaltic boulders and outcrops along the road here.
0.2	20.5	Cross West End Wash. There is a paved turnout on the left side of the road just beyond the wash. The prominent sharp-crested ridges about 1 mile north (left) of the road are the tan to cream-colored limestones of the Bitter Ridge Limestone Member of the Horse Spring Formation. These limestones were formed as calcium carbonate precipitated in a freshwater lake that filled a basin formed during regional extension in Miocene time.
0.1	20.6	Red Aztec Sandstone on both sides of the road. Note the black basaltic boulders littering the red outcrops (see color photo 7c on page 51). The source of the boulders can be traced upward to a thin basalt flow capping the top of the surrounding sandstone hills.
0.4	21.0	Large paved turnout and overlook on the left. The red rock on the right side of the road as you enter the turnout is Aztec Sandstone. The soft, red shaly rocks exposed in the wash to the left of the paved area belong to the Triassic Kayenta, Chinle, and Moenkopi Formations (248 to 206 million years old) that lie below (are older than) the Jurassic Aztec Sandstone (about 180 million years old). You saw these same units back at mileage 3.8 and here, as there, you can see masses of white gypsum weathering from the red sedimentary rocks. Some of these rocks are stratigraphically equivalent to rocks that make up the Painted Desert and Petrified Forest of northeast Arizona. Their vivid colors are a result of a combination of normal weathering processes and a generally oxidizing environment. Yellow colors are produced by hydrous iron oxide minerals, red colors by anhydrous iron oxide minerals (hydrous minerals contain water, anhydrous minerals do not), and black by manganese oxide minerals. Combinations of these minerals produce the various intermediate hues. Like cake coloring, a little goes a long way. A mere trace of a mineral can give strong color to a formation. In sandstones, the color is usually not in the sand grains themselves, but in the cement (silica, iron oxides, or calcium carbonate) between the grains. *(GPS 102)*
0.9	21.9	Paved turnout on the right. Rock exposed here is dark maroon sandstone of the Moenkopi Formation. *(GPS 103)*
0.3	22.2	The ridge in the background at 9:00 is composed of the gray, Virgin Limestone Member of the Moenkopi Formation which contains some gypsum beds.
0.3	22.5	Another paved turnout on the right. From here is a good view of Hamblin Mountain at 1:00. *(GPS 104)*

Photo: Becky Purkey

The sharp-crested ridges of the Bitter Ridge Limestone Member of the Horse Spring Formation.

Photo: Becky Purkey

Hamblin Mountain.

interval | cumulative

0.5 | 23.0

Lovell Wash. On the right (south) of road, notice the large blocks that have fallen into the wash because of undercutting by water during flash floods. Changes in bank configuration are rapid as a result of rare, but powerful, flash floods. *(GPS 105)*

About two miles up the wash to the north is the Anniversary Mine, one of several mines in the Muddy Mountains that produced borate from deposits of colemanite in the upper part of the Horse Spring Formation. In this area, in addition to borate, the Horse Spring Formation contains thin layers of algal mats (fossil organic material deposited through the action of algae), fossil animal tracks, and ripple marks indicative of a mudflat environment.

Photo: Becky Purkey

Large blocks of Muddy Creek Formation in Lovell Wash.

0.3 | 23.3

Cross Callville Wash. The dirt road down Callville Wash (to the right) is very rugged and should only be attempted in a four-wheel-drive vehicle. Be cautious, it is much easier to drive down a sandy wash than it is to return back up the wash. Even with four-wheel-drive it is possible to become very stuck in the sand in an unfamiliar wash. *(GPS 106)*

BORATE DEPOSITS IN THE MUDDY MOUNTAINS

Borate minerals are known to occur in several locations in Nevada and discoveries of "cottonball" borate (ulexite) in the state in 1870 predate discovery of similar deposits in Death Valley, California by about three years.

The largest borate deposits in Nevada are in the Muddy Mountains of Clark County. Deposits have been mined in two separate areas, White Basin in the central part of the Muddy Mountains, and in the area of Lovell Wash in the southern Muddy Mountains. These deposits are found in the Bitter Ridge Limestone Member of the Horse Spring Formation. The borate mineral in these deposits, colemanite (hydrous calcium borate), is thought to have formed in shallow ephemeral playa lakes that occupied lower parts of the extensional basin within which the sediments of the Horse Spring Formation were being deposited. Boron, calcium, and other elements were added to the lake water by hot springs, and the borate minerals precipitated from the water as the playa lakes periodically dried up. The resulting colemanite-rich beds are irregular in thickness, mostly less than three feet thick in White Basin and from about 8 to 18 feet thick in the Lovell Wash area.

Colemanite was discovered in White Basin in 1920 and deposits there were mined and milled by the American Borax Co. until 1924 when litigation with the Pacific Coast Borax Co. halted activity. The Anniversary Mine colemanite deposit, in Lovell Wash about 11 miles southwest of the White Basin deposits, was discovered by F.M. Lovell and G.D. Hartman of St. Thomas in 1921. The deposit was acquired by West End Chemical Co. and production began at the Anniversary Mine in 1921. The operation closed in 1928 due to competition from California mines, and there has been no borate production from the Muddy Mountains, or anywhere in Nevada, since that date. Total production from the Anniversary Mine was about 200,000 tons of borate.

Borax has traditionally been used for household soaps and detergents, but industrial uses, such as heat resistant glass used in kitchen utensils and automobile headlights, now account for most of its consumption.

Ore cars at main adit of the Anniversary Mine, photograph from the company's 1923 Annual Report.

West End Chemical Co. camp and mill at the Anniversary Mine in Lovell Wash, photograph from the company's 1923 Annual Report.

interval	cumulative	
0.3	23.6	The ridge of dark rocks straight ahead is part of the deeply eroded, 12 to 10.5 million-year-old Hamblin-Cleopatra stratovolcano (refer to page 92 for types of volcanoes). This volcano is cut by the left-lateral Hamblin Bay fault which has displaced the northern half of the volcano, Hamblin Mountain, to its present position about 12 miles to the southwest from its southern half, the Cleopatra lobe. (see sketch and map at right). The Cleopatra lobe may be explored ahead at mile 37.4.
0.3	23.9	The low hills to the right of the road are speckled with fragments of selenite.

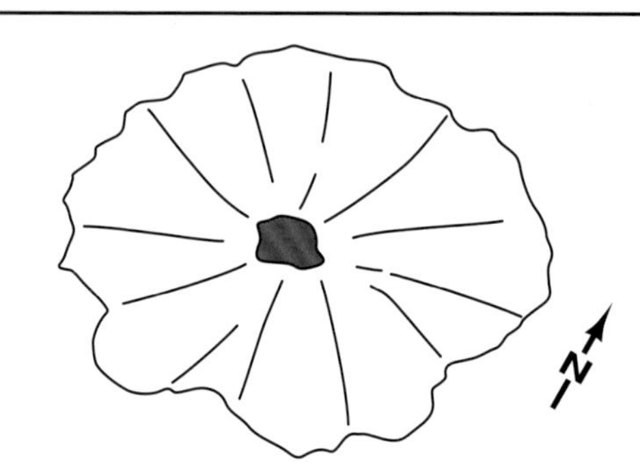

1. The Hamblin-Cleopatra volcano more than 13 million years ago.

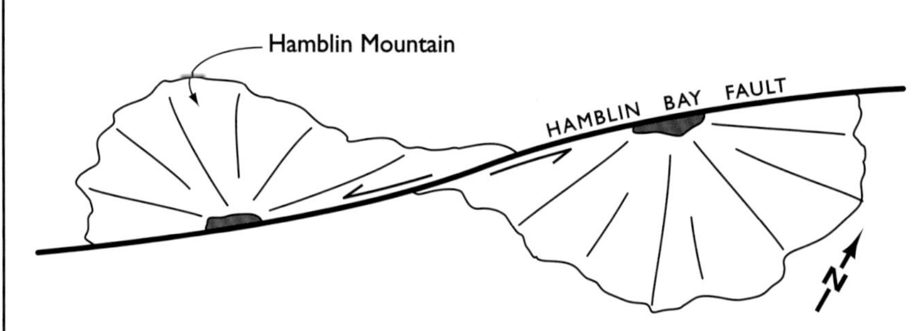

2. The Hamblin-Cleopatra volcano after about 13 million years ago.

Reconstruction of the Hamblin-Cleopatra stratovolcano (after Anderson, 1973; and Bezy, 1978). ▶

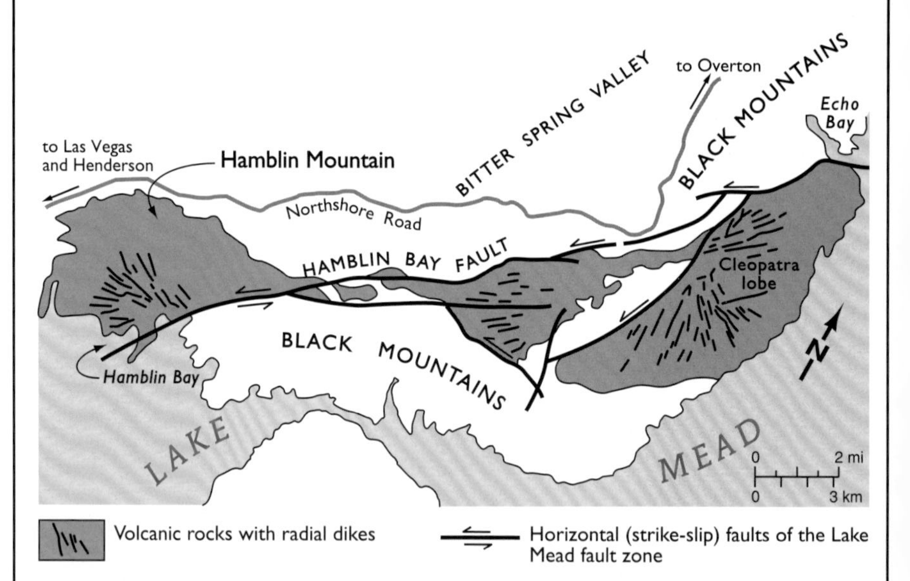

3. The Hamblin-Cleopatra volcano and the faults that dissect it as they appear today.

Aerial view, to the southeast, over the tilted and faulted rocks of the Hamblin-Cleopatra stratovolcano. Hamblin Mountain stands alone in the right foreground.

interval	cumulative	
0.4	24.3	A syncline in the Bitter Ridge Limestone Member of the Horse Spring Formation is quite conspicuous on the left at 10:00. There is a paved turnout on the left here that affords a good view of the syncline, but be careful of oncoming traffic if you decide to cross the highway and stop. *(GPS 107)*
1.1	25.4	Good view of steeply dipping limestone beds on the right. Muddy Peak is visible in the distance at 9:00.
0.1	25.5	Paved turnout on the left. This is a good spot to pull over and view the spectacular talus cones at 1:00. Talus refers to accumulation of angular rock fragments at the base of a slope due to weathering and periodic rock falls. The weathered material is finer toward the top of these steep, very unstable slopes. As the rock debris grows upward, it protects the cliff from further erosion. *(GPS 108)*

Photo: Jack Hursh

A syncline in the Bitter Ridge Limestone Member of the Horse Springs Formation.

Photo: Becky Purkey

Talus cones southeast of Callville Wash.

interval	cumulative	
0.7	26.2	Paved turnout on the left, carefully pull off to view the Bowl of Fire. The Bowl of Fire has formed in the heart of an anticline—an up-arched fold that exposes older rocks at its core. The rocks in the core of the Bowl of Fire anticline are Aztec Sandstone (about 180 million years old). The younger rocks on the flanks include various members of the Horse Spring Formation (17 to about 12.5 million years old) that lie unconformably on and in fault contact with the Aztec Sandstone. The dark rocks in foreground, on each side of the gap leading into the bowl, are the Rainbow Gardens Member of the Horse Spring Formation. *(GPS 109)*

Photo: Becky Purkey

Bowl of Fire.

interval	cumulative	
0.9	27.1	Large paved turnout on the left on the curve. Be very careful if you decide to cross the road here and stop. Northshore Road follows the Lake Mead fault system—a series of faults with left-lateral displacement of up to 40 to 50 miles. Along this fault, rocks of the Horse Spring Formation (deposited about 17 to 12.5 million years ago) on the north side of the road have been displaced west and southwest relative to Permian rocks now located south of the road. One of these faults must pass beneath the parking area and would pass through the low saddle ahead.
0.1	27.2	At 10:00, rock layers roll over into a gentle anticline. Light-colored gypsum-bearing layers are exposed in the core of the fold.

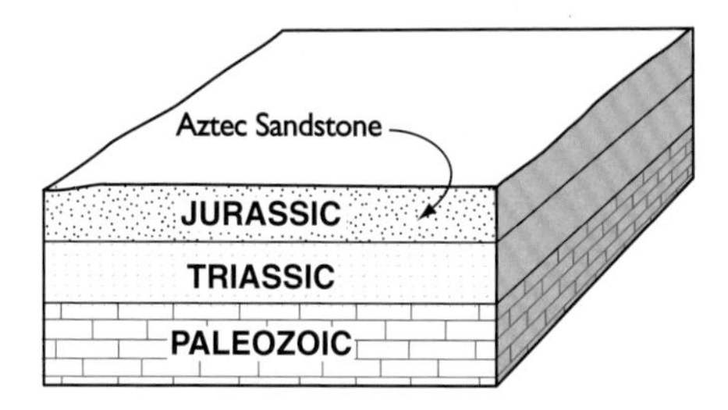

1. Deposition of Jurassic Aztec Sandstone in a vast desert (about 180 million years ago).

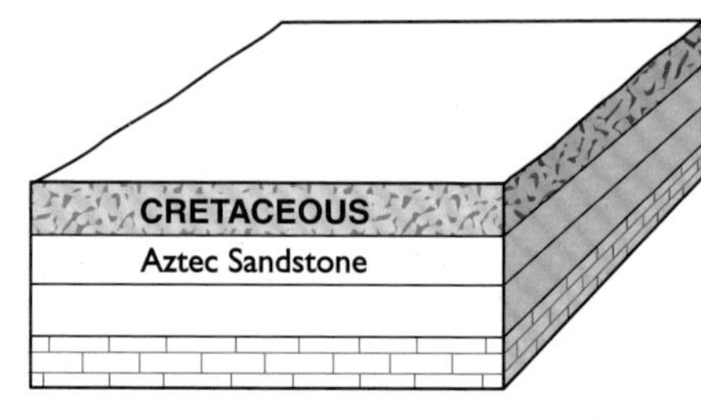

2. Deposition of Cretaceous rocks on top of the Aztec Sandstone.

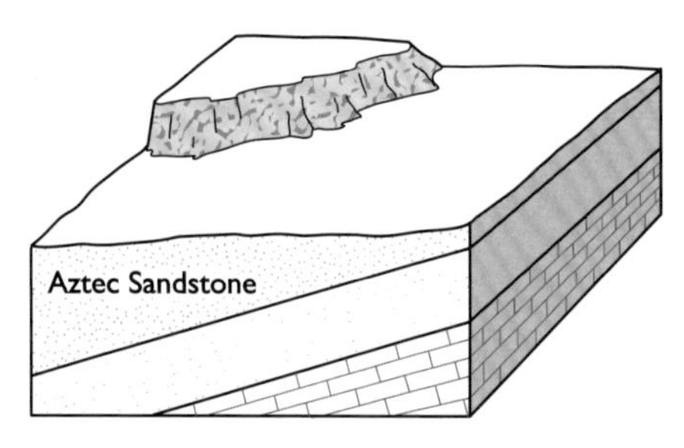

3. Uplift, tilting, and erosion of younger sedimentary rocks expose the Aztec Sandstone.

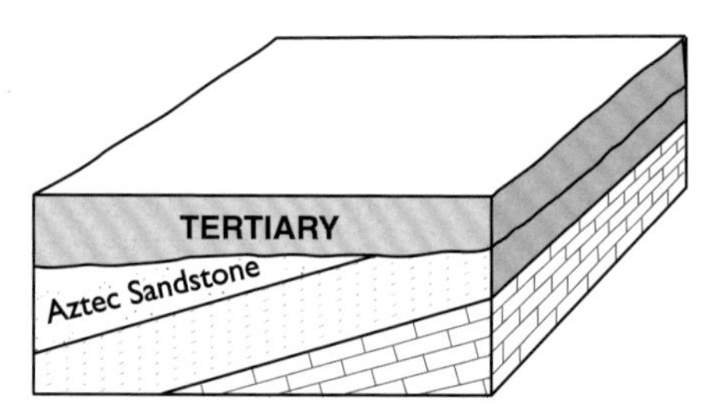

4. Deposition of river and lake sediments once again buries the Aztec Sandstone (about 18 million years ago).

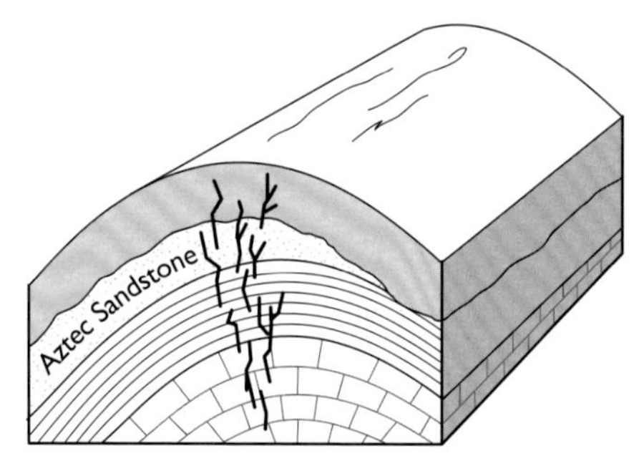

5. Folding of rocks forms an anticline. Folding causes rocks to fracture (about 11 million years ago).

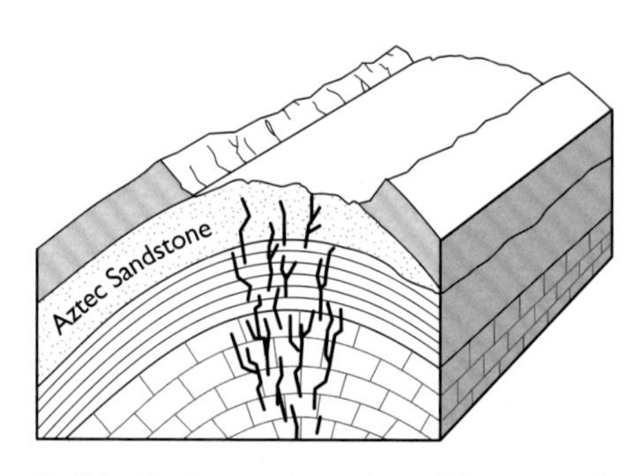

6. Weathering and erosion at the crest of the arch result in exposure of the underlying Aztec Sandstone (present day).

How the Bowl of Fire was formed.

interval	cumulative	
0.7	27.9	Turn off to small loop for access to Northshore Summit Trail on the left. This short (about ¼ mile) and easy hiking trail provides a panoramic view of the entire western Lake Mead area and part of Las Vegas Valley. *(GPS 110)*
		From the parking lot, the trail crosses folded rocks of the Thumb Member of the Horse Spring Formation. Once at the top of the low ridge, you can walk to several points that provide views of the folded red Aztec Sandstone to the west in the Bowl of Fire, of Muddy Peak and its rugged exposures of Paleozoic limestone and dolomite to the northwest, and of the broad expanse of Bitter Spring Valley to the north.
		Return to Northshore Road. Continue eastward and descend into Bitter Spring Valley.
0.7	28.6	If the light is right, you can see the Virgin Mountains in the distance at 11:00. Virgin Peak, the highest point, is 8,075 feet in elevation. The lower flanks of these mountains are composed of Precambrian igneous and metamorphic rocks overlain by Paleozoic sedimentary rocks that make up the high portions of the range.
0.6	29.2	Paved turnout on the right, view of prominent ridge on the right is capped by a resistant conglomerate ledge of the Rainbow Gardens Member of the Horse Spring Formation. The soft, tan and red sediments underlying the ridge form a slope that is now littered with large blocks of the resistant material that have broken from the capping ledge. *(GPS 111)*
0.2	29.4	There is a good view of slump blocks on the right. Notice the gypsum layers and stringers in the red sediments exposed in the road cut.
0.3	29.7	Panoramic view of Bitter Spring Valley at 11:00. The prominent, discontinuous, pink-tan ridge in the distance is composed of the Bitter Ridge Limestone Member of the Horse Spring Formation. This relatively young rock unit (deposited about 13 million years before present) is folded, tilted, and faulted proving that major geologic activity took place in this region after that time. The narrow, rugged buttes protruding from the valley floor are composed of the Thumb Member of the Horse Spring Formation capped by iron-stained Thumb breccia.

OIL IN THE MUDDY MOUNTAINS

Oil and gas can accumulate in geologic structures such as the Bowl of Fire anticline, provided there is a source for the hydrocarbons and the permeable and impermeable rock layers are in a favorable sequence. Shell Oil Co. drilled a well to 5,919 feet in this anticline in 1959 and found the rock sequence to be similar to that in the South Virgin Mountains to the east and in Frenchman Mountain to the west. Oil shows were recorded at 877 feet and at other horizons in Triassic, Permian, and Mississippian strata, but no commercial oil was found.

Photo: David Gray

Panoramic view of Bitter Spring Valley from Lakeshore Road. The buttes are formed from the Thumb Member of the Horse Springs Formation.

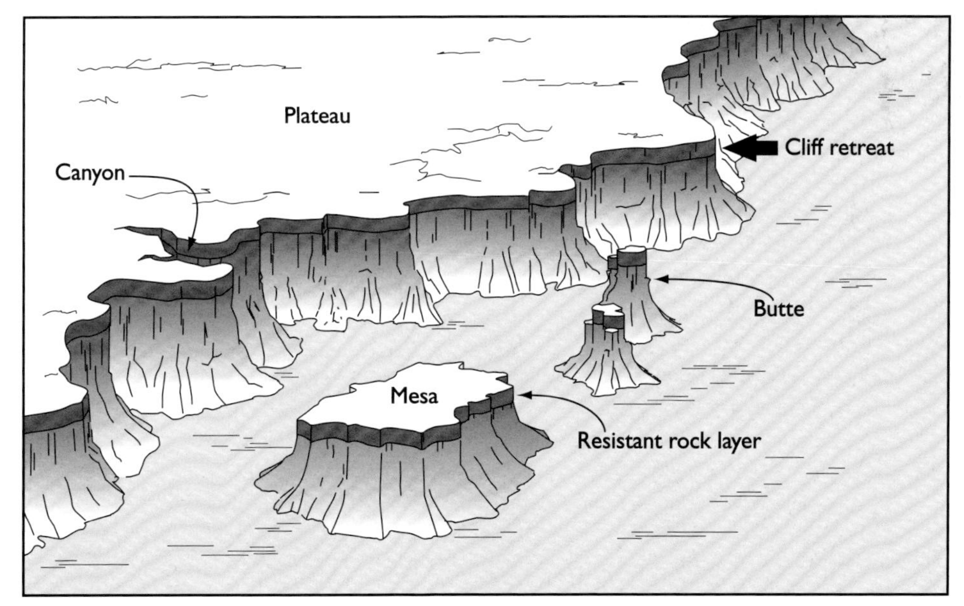

Buttes and mesas form as canyons cut into a cliff of soft sediments capped by more resistant rock.

interval	cumulative	
0.4	30.1	The dark, rough-textured rock in this area (look to the right at 2:00) is limestone of the Permian Kaibab Formation. This rock unit is exposed throughout the southwestern United States (it's the same rock that forms the south rim of the Grand Canyon). Parts of the formation are fossiliferous, and crinoid stem chips, brachiopods, and rugose corals can sometimes be found. We saw this rock at the starting point of this trip, and more outcrops of it are seen on Trip 1.
0.2	30.3	The route now passes through outcrops of the Triassic Moenkopi Formation, composed mainly of reddish-brown, fine-grained siltstone and shale. Fresh surfaces display thin horizontal layering, ripple marks and mud cracks—evidence that this formation was deposited in shallow water in a tidal flat. Many tracks and trails of reptiles that made their home in this tidal flat have been found preserved in these rocks.
0.3	30.6	From the paved turnout here on the left, there is a good view of Bitter Spring Valley and the small buttes that dot the valley. *(GPS 112)*
1.7	32.3	Another paved turnout, but this one has an emergency telephone if you need one. There is a good view from here of red Aztec Sandstone ahead at about 12:00 at the base of the Black Mountains. Ahead, to the left of the section of winding road, are the Echo Hills. Outcrops of Kaibab Formation just across the road to the south of the turnout (the north end of Pinto Ridge) are ribbed black due to rock varnish formed preferentially on some layers. *(GPS 113)*
0.9	33.2	Aztec Sandstone outcrop is straight ahead and extends to the base of the distant, rounded hill. The dark layer on the top of the hill is the Rainbow Gardens Member of the Horse Spring Formation, which lies unconformably on the Aztec Sandstone. The Rainbow Gardens Member is estimated to have been deposited at least 17 million years ago; the Aztec Sandstone was deposited about 180 million years ago. This gap in the geologic record (an unconformity), therefore, represents a time span of some 160 million years. This is the same unconformity seen just east of Frenchman Mountain at mile 5.8.
0.2	33.4	Bitter Spring road is to the left; Echo Hills are at 11:00.
0.4	33.8	On the right at 3:00 note the flat, step-like terraces midway up the ridge. They are remnants of ancient valley floors that were covered with sand, gravel, and cobbles that came from the ridge to the southeast, then were uplifted during Miocene time, and now are being dissected by streams to form a lower alluvial terrace. Geologists have identified two sets of successively lower (younger) terraces in this area, which are evidence of at least two major periods of uplift. The highest flat-topped terrace is all that is left of the original valley surface.

Photo: Carol McKim

Cliff face on the north side of Bitter Spring Valley. The cliff is cut in the Bitter Ridge Limestone Member of the Horse Spring Formation.

Photo: Becky Purkey

Terraces near Pinto Ridge.

interval	cumulative	
0.5	34.3	The Redstone Picnic Area is on the right in the red Aztec Sandstone. There are tables and rest rooms here. See photo 6c on page 50. *(GPS 114)*
0.3	34.6	The jumbled, rugged terrain of the Echo Hills is north of the road here. The hills are capped with limestone of the Kaibab Formation, and you can see outcrops of red Triassic rocks low on the base of the hills.
0.9	35.5	Note the nearly vertical bedding in the rocks on the skyline to the right.
1.2	36.7	View to the northwest through Bitter Spring Valley. The Muddy Mountains are visible in the distance from about 9:00 to 12:00. The high gypsum content of the rocks and soils in Bitter Spring Valley makes it a harsh environment for many plant types, and water produced from springs and wells in these rocks is bitter to the taste, hence the name "Bitter Spring Valley." In autumn, particularly late September through October, watch for migrating tarantulas crossing the road in this area.
0.6	37.3	The Boathouse Cove turnoff is on the right. For a walk through the Cleopatra lobe of the Hamblin-Cleopatra stratovolcano, turn right here and travel about 2 miles south to the intersection with Cleopatra Wash. From that point, a walk down the wash toward Cleopatra Cove will take you through the volcano's eroded interior where dikes of intruded magma form resistant ridges above softer, more easily weathered volcanic debris. Please use caution if you take this side trip. The first two miles are four-wheel-drive road, and the walking part is downhill into a hot, dry wash (with an uphill return!). Carry water with you, and don't attempt the hike in hot weather. *(GPS 115)*
0.5	37.8	Good view, straight ahead, of East Longwell Ridge in the Muddy Mountains. Note the deep V-notch cutting the ridge. This is an hourglass canyon, a feature found in desert climates that we will say more about ahead at mileage 44.6.
2.0	39.8	Badlands topography can be seen on the left in Echo Wash. Soft sediments here are composed largely of clay that weathers to barren, round-topped hills. The white and reddish sediments are early Horse Spring Formation (17 to about 12.5 million years old). Down the wash toward the right, you can see crudely-stratified, cemented gravels of the late Miocene to early Pliocene Muddy Creek Formation (10 to 5 million years old). East Longwell Ridge is in the background beyond Echo Wash.

Photo: George W. Robinson, California Academy of Sciences

The desert Southwest is home to a number of **tarantula** species. These spiders have hairy, brown to black bodies, and a leg span that can exceed 4 inches. Despite their intimidating appearance, they are generally non-aggressive and will avoid confrontation if possible. North American desert tarantulas do have venom, but the effect on a human is generally equivalent to a bee sting.

For most of the year tarantulas are solitary creatures that seldom move far from their burrows to hunt lizards, insects, and other arthropods. They are most likely to be seen on summer nights when males are out and about searching for mates. The mortality rate for mature males is high. Those who manage to survive the search for a female will most likely be eaten by her soon after mating. Females, who tend to stick close to the home burrow, may live for 20 years or more.

Photo: Joe Tingley

Barrel cactus.
(*Ferocactus acanthodes*)

interval	cumulative	
0.9	40.7	Cross Echo Wash. The ridge on the left, extending from about 10:00 to 12:00, is East Longwell Ridge, named for pioneering geologist Chester Longwell. *(GPS 116)*

Chester Longwell produced some of the earliest geologic maps of southern Nevada and was among the first to recognize major geologic structures such as the Las Vegas Valley shear zone and the Keystone-Muddy Mountains thrust fault. His record of publication spanned six decades—from the 1920s to the 1970s—a remarkable career.

The Longwell Ridges (there are two; West Longwell Ridge is hidden behind East Longwell Ridge) are made up of Paleozoic limestone and dolomite similar to rocks present in the Spring Mountains west of Las Vegas. In geological terms, Paleozoic rocks in the Muddy Mountains, such as those in the Longwell Ridges, are more similar to those in the Spring Mountains many miles to the west than they are to the rocks immediately south of Northshore Road. Major strike-slip faults of the Lake Mead fault system form the boundary between these two different geologic blocks.

The carbonate rocks exposed in the Longwell Ridges, and to the northeast in the eastern Muddy Mountains, terminate abruptly here due to the presence of the major northeast-trending Rogers Spring fault. The trace of the Rogers Spring fault is marked by a line of discontinuous springs along the cliffs to the left of the road (see sketch at right). Springs occur here because movement along the fault has placed permeable rocks against impermeable rocks. Groundwater that easily flows through permeable fractures in the limestone and dolomite encounters less permeable basin fill at the fault. The basin fill acts like a dam, and water discharges from low points along the fault trace. The shallow groundwater, which reaches the surface in a few springs, is capable of supporting vegetation.

interval	cumulative	
2.0	42.7	The turnoff to Echo Bay Road is on the right. Services available at Echo Bay include lodging, camping, gas, food, and boat rentals. The light-colored rock exposed for about the next mile on the right side of the road is called caliche, a naturally occurring calcium carbonate cemented rock that typically develops in arid climates. *(GPS 117)*

Caliche

Soils in arid climates are typically high in alkaline minerals because of the lack of rain and vegetation, both of which, in chemical interaction with the atmosphere and moisture, normally provide weak acids to soil. In addition, the limestone mountain ranges such as the Muddy Mountains provide abundant calcium carbonate, which is leached out of the rock and deposited downstream on alluvial fans and other desert surfaces.

Moisture evaporates so quickly here that the dissolved calcium carbonate precipitates as crusts on pebbles on the surface and several inches or feet down into the subsoil. (In a wetter climate, the minerals would be washed away more effectively.) In time, the spaces between the individual pebbles in the subsoil will actually fill with the calcium carbonate cement, forming a new sedimentary rock, caliche.

Caliche may vary in color from reddish brown (which means that iron oxide is present in the cement) to white. Silica may also be present with the calcium carbonate to form a particularly hard cement.

Caliche deposits along Echo Bay Road. *Photo: Becky Purkey*

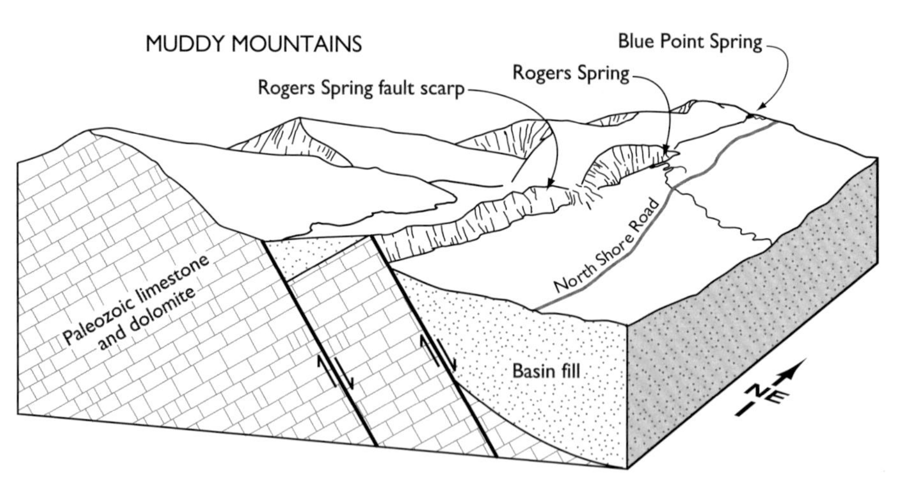

The Rogers Spring fault (after Bezy, 1978).

interval cumulative

1.9 **44.6** Paved turnout. The rugged, steep canyons on the left are typical of those formed in desert climates prone to flash flooding. They are called "hourglass" canyons because of their wide tops and mouths and narrow waists. Fast-moving water flowing from the canyon during intermittent flash flooding is cutting away older alluvial fan gravels deposited at the mouth of the canyon. *(GPS 118)*

0.4 **45.0** At 10:00, a cut bank in the wash exposes multicolored rocks. This exposure marks the trace of the Rogers Spring fault where resistant carbonate rocks of the eastern Muddy Mountains are placed against easily eroded rocks of the Horse Spring Formation. A walk up the bank will reveal the evidence of faulting between the two rock units. You can see altered, discolored, bedded rocks in the footwall of the fault, then really chewed-up rocks in the drag zone. Gravel deposits hide the hanging-wall rocks.

There is no good place to stop and park here so it may be best to observe the color flash in the bank of the wash from your car, and keep driving.

The Overton Arm of Lake Mead is straight ahead. Overton Arm is the valley of the Virgin River that was drowned when Lake Mead filled behind Hoover Dam. The Virgin River once joined the Colorado River just a few miles south of here.

0.3 **45.3** If you want, pull over on the wide shoulder here for a panoramic view of some features in the South Virgin Mountains. Look across Lake Mead at about 1:30. Lime Ridge, to the north, is separated from Anderson Ridge, to the south, by a deep V-notch. Through the deep V-notch in the mountains, the point of Gold Butte (elevation 5,042 feet) can be seen in the background. The old townsite of Gold Butte (not visible) is in the valley to the left of the point. Mica Peak (elevation 5,758 feet) is at about 2:00, and Bonelli Peak (elevation 5,334 feet) is at about 2:45, just visible over the alluvial ridge to the south of the highway. Gold Butte is a major player in the Frenchman Mountain story told earlier. *(GPS 119)*

Photo: Joe Tingley

V-notch marking an hourglass canyon west of Rogers Spring in the eastern Muddy Mountains.

Photo: Joe Tingley

The Rogers Spring fault exposed in the steep bank of the wash. The fault trace generally runs parallel to the wash here. The limestone in the rugged mountain face on the left is in the footwall of the fault, the crushed rock marks the fault trace, and the gravel on the right is in the hanging wall of the fault.

interval	cumulative	
2.0	47.3	The turnoff to Rogers Spring Picnic Area is on the left (see color photo 9e on page 117). Turn in and park. The spring flows year round at a high rate and a constant temperature of about 80°F and hosts many (although non-native) fish including perch, mollies, and golden shiners (look for the small turtles who also like the warm water). Vegetation at this oasis includes tamarisk, saltbush, cattails, and palm trees. *(GPS 120)*
0.4	47.7	Leave Rogers Spring area. Turn left onto Northshore Road.
0.6	48.3	A few small cottonwood trees are present in the wash to the right of the road. These are the first trees you have seen since leaving Las Vegas. Other vegetation in this area consists mainly of blackbrush and creosote bush.
0.1	48.4	Blue Point Spring, another of the small springs along the Rogers Spring fault, is on the left. *(GPS 121)*
0.4	48.8	Turnoff to Stewarts Point on the right. The white, rough-textured deposit on either side of the road is gypsum of the Muddy Creek Formation (10 to 5 million years old). The gypsum was deposited in a playa—a lake bed that is usually dry but is subject to periodic flooding. The floodwater usually evaporates in a few days, leaving deposits of soluble salts such as gypsum and halite (common rock salt). *(GPS 122)*
1.0	49.8	Crossing Valley of Fire Wash. The large, dark green (in summer) treelike plants in the bottom of wash are mesquite. *(GPS 123)*

Photo: Joe Tingley

Mesquite.

GOLD BUTTE

The rock outcrops in the South Virgin Mountains between Gold Butte and Bonelli Peak are Precambrian gneiss and rapakivi granite, presumed source rock for the landslide breccia units in the Thumb Member of the Horse Spring Formation. You are now looking toward the presumed original home of Frenchman Mountain, where geologists believe it was located before it began its westward journey along the Lake Mead fault system and the Las Vegas Valley shear zone some 15 million years ago (see discussion on page 58). The Gold Butte fault, occupying the valley between Lime Ridge and the mountains to the south, is a Lake Mead fault system structure, and Frenchman Mountain would probably have been originally located north (to your left) of the fault.

The name Gold Butte is taken from the ghost mining camp of that name that is located at the base of the peak. Daniel Bonelli, an early Mormon settler in the Virgin Valley, prospected in this area and in 1872 found a deposit of sheet mica near what is now Bonelli Peak, but little came of the discovery. The town of Gold Butte sprang up around 1905 following nearby discoveries of small gold-bearing veins. Another small-scale boom developed when more gold, and also copper, was found in 1908. The deposits were small and the town eventually faded away, but gold was still being mined at the Lakeshore Mine as late as 1941, and a few gold placer operations have been active in recent years.

MESQUITE

A member of the legume family, mesquite is the most common shrub (or small tree) of the desert southwest. The plants have characteristic bean pods; the beans were a staple food of the local American Indians and the beans, pods and all, are a favorite food of cattle. A phreatophyte, mesquite plants have roots that may penetrate to depths of 50 feet or more in search of subsurface water.

interval	cumulative
0.6	50.4

The colorful bedded rocks on both sides of the road here are Muddy Creek Formation capped, in places, by recent gravel deposits. To the right, a 4-wheel-drive road leads to Fire Cove and Salt Cove on the shore of Lake Mead. *(GPS 124)*

At Salt Cove, now only reachable by boat, you can examine outcrops of salt that formed during the past 10 million years in playa lakes, and were then buried under hundreds of feet of younger sediments. Because salt is lighter (less dense) than the rocks burying it, it migrated to surface as a plastic mass under the pressure of burial and tectonic forces that shaped the present topography. The resulting feature is called a salt dome, and several small salt domes are exposed along the edge of Lake Mead in this area.

1.6 | 52.0

Mormon Mesa is straight ahead, bounded by the valley of the Virgin River on its right and the valley of the Muddy River (Moapa Valley) on its left. The upper reaches of the Overton Arm of Lake Mead extend a short distance up each valley. The red rocks to the left in mid-distance are within Valley of Fire State Park.

Mormon Mesa, view to the northeast. *Photo: Becky Purkey*

0.6 | 52.6

The turnoff to Overton Beach is on the right. Facilities and accommodations for fishermen can be found at the beach. Favorite fish sought here are catfish, largemouth bass, and rainbow trout. *(GPS 125)*

As you continue ahead, the road curves to the left and begins a gentle climb toward the Valley of Fire turnoff. In this area, notice the white, sparkling rocks exposed on both sides of the highway. This is a gypsum-rich horizon within the Muddy Creek Formation.

Salt Deposits in Virgin Valley

Thick beds of naturally occurring rock salt occur in the lower part of the late Miocene to early Pliocene Muddy Creek Formation (10 to 5 million years old) in this area along the Virgin River. The salt beds were exposed in cliffs at numerous places for about 12 miles along the Virgin River Valley south of the old town of St. Thomas. With the exception of the small salt domes near Salt Cove, all of the salt bed outcrops, as well as the site of St. Thomas, were covered by Lake Mead. With the current (2008) low water level of the lake, the salt beds and the ruined foundations of the town are again above water.

Virgin Valley salt was mined by American Indians who lived in the valley from about 700 to 1150 A.D., and they may have traded this commodity as far south as Mexico. Their principal source was a natural cave in what was described as "a mountain of salt" about four miles south of St. Thomas. The salt was mined from the walls and floor of the cave by chipping circular channels and undercutting to remove discs several feet in diameter. A few stone hammers found in the cave by archaeologists may now be seen in the Lost City Museum in Overton. The museum also has a small exhibit showing a drawing of the salt cave, and even some salt on display.

The salt cave was seen by fur trapper Jedediah Smith when he passed this way in 1826, and J.S. Newberry, geologist with Lt. Joseph Ives' Colorado River expedition of 1857–1858, was told of the salt deposits and reported "On the banks of a tributary of the Rio Virgen, a few miles from the Colorado, the Indians obtain large quantities of rock salt. Much of it is very pure and beautifully crystallized, and it is said to exist in immense quantities." Early white settlers in the Virgin Valley mined the salt to a limited extent for local human and livestock consumption, and thin slabs of the clear, crystalline salt were even tried as window panes. Starting about 1866, salt was mined from the deposits and shipped to silver chlorination mills in the Eldorado Canyon district (see Trip 4, page 109) and also to the Mineral Park district in Arizona.

Although the salt deposits have been described as "virtually inexhaustible," very little was ever produced from them because of transportation difficulties and distance from large markets. About 200 tons was mined annually from the Virgin Valley deposits from the 1920s until the mines were flooded by the rising water of Lake Mead in 1937 to 1938.

Photo: Nevada Historical Society

Ruins of the U.S. Post Office building in St. Thomas reflected in the rising waters of the Overton Arm of Lake Mead, 1938.

interval	cumulative	
1.4	54.0	The turnoff to the Valley of Fire State Park is on the left; the road to St. Thomas Cove (four-wheel-drive only) is on the right. The site of historical St. Thomas, now exposed by the receding waters of Lake Mead, is located about 3/4 mile beyond the end of this road. *(GPS 126)*
		Reset odometer to 0.0 and continue straight ahead for an optional side trip to the town of Overton (all services) and the Lost City Museum, or turn left onto Valley of Fire Road. The 18-mile round trip to Overton is well worth the time, if only to visit the museum.
	0.0	Reset odometer to 0.0 when you turn onto Valley of Fire Road to begin the final segment of this trip. *(GPS 127)*

OVERTON SIDE TRIP

interval	cumulative	
0.1	0.1	For the next two miles, the route crosses through an area of wind-blown sand, eroded from the Aztec Sandstone bluffs to the west. This is a present-day example of how sand dunes form and migrate, grain by grain (refer to Trip 1, page 27, for more information on dune formation).
1.4	1.5	Leave Lake Mead National Recreation Area; the road now becomes Nevada S.R. 169. Between 1:00 and 3:00 is Mormon Mesa. The mesa is capped by a resistant deposit of caliche (see photo, page 79). *(GPS 128)*
1.0	2.5	Ahead, steeply east-tilted cemented conglomerate beds of the Horse Spring Formation are seen on both sides of the road. These conglomerates cap Overton Ridge, which extends from here to the northwest for several miles.
0.2	2.7	To the left of the highway, west of Overton Ridge, are two open-pit silica mines operated by Simplot Industries. You can see one pit from the road, the second is about one mile farther to the northwest. The pits are dug in a large deposit of friable sandstone (easily crumbles into sand), the Cretaceous Baseline Sandstone. These are active mines and are not open to the public.
0.5	3.2	The Grand Wash Cliffs form the western face of the flat mesa on the skyline in the far distance at 3:00 (between the higher mountains). The Grand Wash Cliffs mark the topographic break between the Colorado Plateau and the Basin and Range provinces. The Basin and Range has been subject to major deformational events in the Mesozoic and Cenozoic Eras (discussed in the Introduction), whereas the Colorado Plateau has been deformed to a much lesser degree.

Pueblo foundations at the Lost City Museum (the Pueblo Grande de Nevada), Overton. ▶

A display at the Lost City Museum showing the progression of an archaeological dig from an undisturbed site (foreground) through careful excavation to the fully excavated site (background). ▼

Photos: Mark Vollmer

interval	cumulative	
0.2	3.4	The flat, open area on both sides of the highway here at the top of the hill is likely to be populated with RV's of all types—trailers, motor homes, and pickup campers. In the winter the numbers will be high, and there may even be one or two here in the hottest months. It is apparently a good overnight spot for snowbirds heading north or south, and the scenery and local attractions make it a good longer-term camp for some who make this their winter destination. This flat has formed on a cemented gravel layer in the Muddy Creek Formation. To the left, Kaolin Wash has cut through this resistant layer and has proceeded to wash away the softer layers beneath, forming badlands extending to the north. If you stop and walk to the edge of the flat, you can see evidence of this undercutting. Looking across the wash to the north, you see another flat-topped bench, capped by the same cemented gravel layer.
0.2	3.6	Begin descent into Moapa Valley. Kaolin Wash is to the left.
2.0	5.6	The flat-lying deposits along both sides of the road belong to the Muddy Creek Formation. Here the Muddy Creek Formation is more silty and you do not see the cemented gravel layers that cap the motor home parking lot back at mileage 3.4.
0.2	5.8	Pistachio orchard on right. *(GPS 129)*
0.1	5.9	The road crosses another wash. Note the large tamarisk trees on the left as well as some mesquite trees in the wash to the right.
0.7	6.6	Crossing Magnesite Wash. East Waterfowl Road, leading to the Overton Wildlife Management Area, is to the right. The Simplot Industries silica processing plant is on the left. *(GPS 130)*
0.1	6.7	Pueblo Grande de Nevada historical marker is on the right. *(GPS 131)*
0.3	7.0	Magnesite Road to the left. About 1.5 to 2 miles down this road is an area of magnesite deposits that occur in a limestone unit in the Rainbow Gardens Member of the Horse Spring Formation. *(GPS 132)*
0.7	7.7	Entering Overton. Many early explorers passed through here on their travels to and from California. The names of Jedediah Smith, Kit Carson, John Frémont, and Jacob Hamblin are firmly established in the history of this area. The name Overton may be derived from the term "over town" which is how the inhabitants of the Mormon Hill settlement, established on the east side of the Virgin River, referred to the newer town over the river. The present site of Overton was selected in 1880. Formerly part of Lincoln County, this area was transferred into Clark County when the new county was created in 1909. Many historical houses are still standing in the town.
0.2	7.9	Entrance to the Lost City Museum is on the left. *(GPS 133)*
0.5	8.4	At last, a McDonald's!

Silica and Magnesite Deposits at Overton

Sand is mined from the Cretaceous Baseline Sandstone about 4 miles southwest of Overton (see comment at mileage 2.7). The sandstone contains nearly 97 percent silica. It is washed at the mine site and piped as a slurry to the plant at Overton where it is dried, screened, and shipped in bags or as a bulk product. The final product contains more than 99 percent silica, making it highly desirable for use in production of glass and in the chemical industry.

There are also extensive reserves of magnesite (magnesium carbonate) in this same general area, but it occurs in fairly thin beds interlayered with white, clayey dolomite. Lime and silica impurities must be separated and removed from the magnesite to obtain a commercial product, and its fine grain size makes this an expensive process. Only small amounts of magnesite have been produced from these deposits, and they are today idle. The major uses for magnesite in the United States include: the manufacture of magnesite bricks for use in metal, cement, and glass production furnaces; as an additive in animal feeds and fertilizers; and in the chemical processing of manufactured products such as rayon, fuel additives, rubber, pulp and paper, pharmaceuticals, and sugar.

Lost City Museum

This museum contains an extensive collection of local American Indian artifacts with material from the Desert Culture of 10,000 years ago, the Basketmaker Culture that lasted until about 500 A.D., the Pueblo culture from 500 to 1150 A.D., and the Paiutes who entered the area around 1000 A.D. The museum preserves the story of Pueblo Grande de Nevada (the Lost City) which was situated near the base of Mormon Mesa on the east side of the Virgin River Valley (now the shores of Lake Mead). Scores of villages, scattered along a distance of almost 30 miles, occupied the valley from before the time of Christ to about 1150 A.D. The latest inhabitants, the Puebloans, raised maize and cotton, mined salt and turquoise, and made fine pottery. Fay and John Perkins of St. Thomas "discovered" the sites in the early 20th century and excavations of the Lost City were begun in the 1920s.

Lost City Museum outdoor exhibit.

Photo: Becky Purkey

interval	cumulative	
0.2	8.6	This ends the optional trip to Overton. Turn right here about one block to Overton Park, a nice spot for a picnic, or turn around and retrace your route on S.R. 169 to the Lake Mead National Recreation Area boundary, then continue to the Valley of Fire Road turnoff to resume your trip to Valley of Fire State Park. *(GPS 134)*
	0.0	Turn right into Valley of Fire State Park and reset your odometer to 0.0. *(GPS 135)*
0.2	0.2	The Muddy Mountains can be seen on the skyline, from 10:00 all the way to the right. Ahead low on the mountain flank, you can see a small red flash. This is Fire Alcove, one of the features within Valley of Fire State Park.
0.9	1.1	Enter the Valley of Fire State Park. This is Nevada's first state park, created in 1923 under the leadership of then-Governor James G. Scrugham. The roads are narrow within the park, so be careful, take your time, and enjoy the park. *(GPS 136)*
		At this point, we pass from Muddy Creek Formation, exposed on the left, into red Aztec Sandstone, on the right. Many of Valley of Fire's most spectacular features are carved in Aztec Sandstone.
0.6	1.7	The prominent white dome with its top sloping to the east visible through the notch at about 2:00 is Silica Dome. The dome is composed of Cretaceous Baseline Sandstone, the same formation that hosts the silica mines west of Overton Ridge.
0.3	2.0	Elephant Rock, stop here and pay the park entrance fee. While you are here, take time to read the informational signs. *(GPS 137)*
0.3	2.3	Old Arrowhead Trail historical marker on the left. This is a remnant of an early automobile road built in 1915 between Salt Lake City and Los Angeles. The cliffs on the right (north) side of road are composed of Aztec Sandstone. The rocks exposed in the low ground in front of the cliffs are part of the Triassic Chinle Formation. *(GPS 138)*
		The Chinle Formation consists of a thick sequence of shale which displays a variety of brilliant colors ranging from blue, purple, pink, green, gray, maroon, to brown. Because shale is relatively soft and very fine grained, it weathers into slopes, gullies, low dome-shaped hills, and badlands.
		Fire Alcove is now visible across the valley to the left at 11:00. From here there is a good view of the thrust fault contact between the Aztec Sandstone and the older limestone and dolomite above the thrust. All the gray rocks are in the upper plate, all of the red rocks are in the lower plate.

VALLEY OF FIRE

The geology of the Valley of Fire is very similar to that of the Red Rock Canyon area west of Las Vegas (Trip 1). In the Red Rock area, the dominant geological feature is the Keystone-Wilson Cliffs thrust fault system that placed older Paleozoic limestone and dolomite on top of younger Jurassic Aztec Sandstone. The same situation, older Paleozoic limestone and dolomite thrust over Jurassic Aztec Sandstone (with some additional geologic complications), exists in the Valley of Fire. The main thrust fault in this area is the Muddy Mountains thrust. Because of these similarities, geologists have correlated these thrust faults and suggested that they are remnants of a formerly more extensive single thrust sheet that was dismembered by movement along the Las Vegas Valley shear zone. The geologic complication is there may be more than one thrust sheet exposed in the Valley of Fire. The red Aztec Sandstone is in the footwall of an older thrust fault, and is overridden by Permian rocks. These rocks are, in turn, overridden by more Permian rocks. In another complication, the Arrowhead fault, the fault separating gray rocks from red rocks along the south side of Valley of Fire, is thought to be one and the same as the Muddy Mountains thrust fault to the west, but to the east end of the valley, on each side of Fire Alcove, the Arrowhead is a Tertiary high-angle fault. We will not concern ourselves with these complications, keep the gray limestone and dolomite in your mind as upper plate and the red sandstone as the prime lower plate rock, and let the geologists worry about finding the correct fault.

Photo: Mark Vollmer

White sand eroding from the Cretaceous Baseline Sandstone (low rounded hill in the background through the notch) fills a wash cut through a jagged ridge of Aztec Sandstone, Valley of Fire State Park.

interval	cumulative	
0.3	2.6	John H. Clark historical marker on the left (a white cross in the wash). *(GPS 139)*
		The hills in this area are covered with the silver gray desert holly and there are some large mesquite in the wash to the right. There is a striking variety of plant life that has adapted to this area. Hardy creosote bush, bursage, and brittlebush are known for their ability to survive in the hot desert. Another group commonly found farther north or at higher elevations because of its ability to adapt to cold conditions includes purple sage, blackbrush, and banana yucca. Still other species, such as Spanish bayonet and Mormon tea, have adapted to cold climates and the sandy soils produced by the weathering of the Aztec Sandstone.
		During the early spring (April and early May), except in the driest years, wildflowers such as yellow desert marigold, orange Indian paintbrush, and the small, dark flowers of indigo bush add special beauty to the face of this otherwise harsh country.
0.5	3.1	Petrified log on the right. To view this specimen, take the path up the hill beyond the parking area. *(GPS 140)*
0.3	3.4	The Cabins Picnic Area is to the right. These cabins were built in the 1930s by the Civilian Conservation Corps (CCC) and originally served as shelter for people traveling through the area. *(GPS 141)*
1.2	4.6	Seven Sisters Picnic Area. Brilliant red outcrops of Aztec Sandstone occur in the foreground on the left. The ridges behind these patches are composed of dark gray Paleozoic limestone. Here, these rock units are separated by the Arrowhead fault. *(GPS 142)*
0.7	5.3	Turn off to RRCNRA Visitor Center and Mouse's Tank. The Visitor Center provides rest rooms, drinking water, exhibits on local geology and natural history, and a store well stocked with books and works of local artists. The short, optional drive to Mouse's Tank takes you through a narrow canyon to a picnic area and trailhead with access to an easy hike along a wash lined with magnificent petroglyphs. *(GPS 143)*
0.1	5.4	At the "Y" in the road, continue ahead another 0.1 mile to the Visitor Center, or turn left and drive about one mile to reach the Mouse's Tank picnic area. You can continue north beyond the picnic area on the Mouse's Tank road for more views of geology and spectacular scenery. *(GPS 144, 145)*
		A special note to "Star Trek" fans, the Mouse's Tank area served as the set for the final scenes of the movie "Star Trek—Generations."
		When you are finished exploring this area, return to Valley of Fire Road and reset your odometer to 0.0.
	0.0	At the intersection, Visitor Center Road and Valley of Fire Road, turn right (west) and continue through the park. *(GPS 146)*
0.4	0.4	In addition to the characteristic cross-bedding patterns, two other features, differential weathering and rock varnish, typical of the Aztec Sandstone are visible at 3:00. Differential weathering is discussed on page 84.

Comments on Petrified Wood

Several layers within the Chinle Formation contain petrified logs. The presence of petrified logs indicates that, at the time the Chinle was deposited in very quiet waters, extensive forests were present in this region. Some logs are several feet in diameter. This is the same rock layer that makes up the Petrified Forest in northeastern Arizona.

Petrified wood forms when fallen trees are buried rapidly by sediment. Rapid burial prevents decay of the trees. Groundwater that contains dissolved silica (usually derived from the weathering of volcanic ash) interacts with the cellulose of the wood and replaces it with very finely crystalline silica. Because the replacement process occurs on a molecular scale, the original patterns of the wood are faithfully preserved even though the composition of the tree or log has been completely changed from cellulose to silica.

Please do not disturb any petrified wood you might come across in the park, or elsewhere for that matter. Observe this unique material in its natural state and leave it for future visitors to study and enjoy.

Photo: Mark Vollmer

Log of petrified wood, Valley of Fire State Park.

Photo: Becky Purkey

The Cabins.

interval	cumulative	
1.5	1.9	Paved road to Atlatl Rock and Arch campgrounds is on the right. The petrified wood walk, about a one mile round trip, is on the left. Atlatl Rock is another site of spectacular Indian petroglyphs. *(GPS 147)*
0.7	2.6	The Beehives. An outcrop area of Aztec Sandstone with lots of small grottos and cavities resulting from differential weathering. *(GPS 148)*
1.1	3.7	State Park exit station, leaving the Valley of Fire State Park. Proceed westward toward Interstate 15. *(GPS 149)*
		At the exit station and for a few tenths of a mile beyond, the road passes through Permian Kaibab Formation. This rock is a fault-bounded sliver within the major thrust fault zone; the lowest thrust fault contact between the thrust slices and the lower plate Aztec Sandstone is buried beneath thin gravel cover in the wash to the right.
0.8	4.5	The route passes through Permian red beds. This entire sequence of rocks has been overturned (folded or tilted to angles greater than 90 degrees). As a result, the older red beds lie on top of the younger Kaibab Formation at this locality. Overturning of rock strata is common in regions of thrust faulting.
		As the road curves sharply to the right and begins a steep climb, note the flat layer of cemented alluvial gravel on the skyline at the head of the wash. This is an old alluvial fan that is now being destroyed by the advancing wash on the right.
0.4	4.9	A good view ahead of steeply dipping red beds with the cemented gravel layer lying across the top of them. This is an angular unconformity, flat-lying Quaternary fan gravel deposited on an eroded surface of tilted Permian red beds (see figure on page 56 for a sketch of an unconformity).
0.4	5.3	You have now reached the top of the old alluvial fan. Look to the left and you can see the fan sloping up toward its source canyon. The broad, flat surface over which you are driving is made of coalescing fans (a bajada) from canyons on both the right and left, that has formed a broad surface that slopes gently to the west. As you have seen, this surface is being eaten away by the canyon draining east into Valley of Fire. *(GPS 150)*

DIFFERENTIAL WEATHERING AND SWISS CHEESE

A characteristic of Aztec Sandstone outcrops that is well illustrated in this area is the occurrence of numerous holes in the rock that impart an appearance similar to Swiss cheese. Sandstone is nothing more than sand grains held together by mineral cement, usually calcite or silica. In this natural process, some parts of the rocks are more tightly and completely cemented. These parts are more resistant to weathering and erosion than other parts. The incompletely or poorly cemented areas are more vulnerable to weathering and erosion processes and therefore wear away more readily creating the conspicuous holes in the sandstone. This process is called differential weathering.

Photo: Patti Gray

Weathered Aztec Sandstone in Valley of Fire State Park.

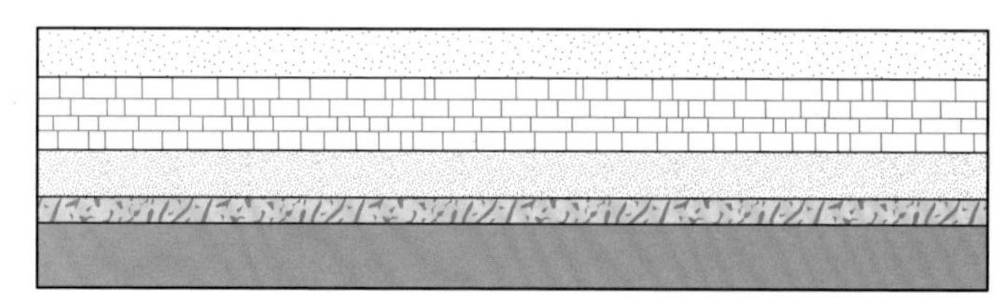

1. Undeformed sedimentary strata prior to thrust faulting.

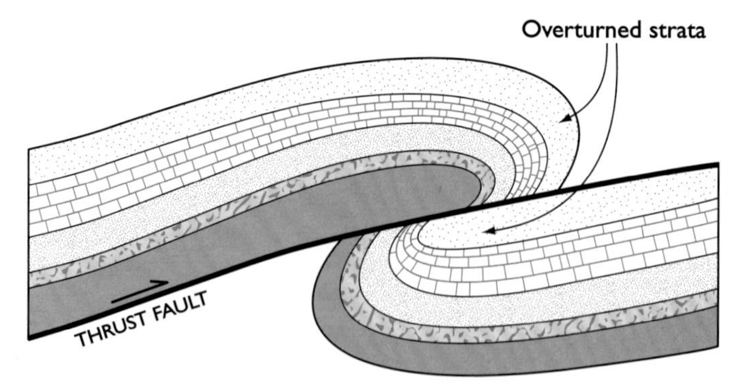

2. Folding and overturning of strata due to thrust faulting.

interval	cumulative	
0.4	5.7	The white and pink sandstone outcrops to the left of the road contain numerous large iron-oxide-stained concretions (spherical masses that are harder than the surrounding rock, see comments on Trip 1, page 24).
1.7	7.4	The well-stratified, variably colored limestones on Piute Point at 9:00 are part of the Permian-Pennsylvanian Bird Spring Formation. Many varieties of fossils can be found in this formation, but it is a fair distance to the mountain front and the best collecting. However, some of the rocks washing down from the mountain contain fossils and a short walk could repay you in fossils and, in addition, introduce you to some of the varieties of cacti native to this area. *(GPS 151)*
1.7	9.1	The Arrow Canyon Range is straight ahead in the middle distance. This range is composed mainly of Paleozoic limestone and dolomite. The high range in the background is the Las Vegas Range.
2.5	11.6	The Buffington Pockets area can be seen to the southeast at 8:00. In this area, the Aztec Sandstone emerges from beneath the Muddy Mountains thrust fault. This thrust-faulting placed older Paleozoic limestone over younger Jurassic Aztec Sandstone. The Buffington Pockets and nearby Colorock Quarry areas were the sites of exploration drilling for petroleum in 1983, although no petroleum was encountered.
6.6	18.2	Intersection with Interstate 15. Continue under the freeway overpass to Las Vegas via I-15 South. *(GPS 152)*
5.8	24.0	A playa occupies Dry Lake Valley at 3:00, the Arrow Canyon Range is in the background.
5.5	29.5	Pass the exit for U.S. 93 north, the "Great Basin Highway," which you would travel to visit Great Basin National Park. The park is a 5 to 6 hour drive to the north from here. *(GPS 153)*
1.7	31.2	Chemical Lime Co.'s Apex lime plant is on the right. High grade lime is produced from pure Devonian limestone by roasting in large rotary kilns. Lime is used extensively in the production of masonry mortar and other building materials, as a flux in steel-making, as an acid-neutralizing substance in the processing of gold ore, in water treatment, and in glass manufacture, to name only a few uses. Some of the lime from this plant is processed in Chemical Lime's plant at Henderson, south of Las Vegas, to make calcium hydroxide (see Trip 4).

Rock Varnish

The black, lustrous coating on many rock surfaces is called rock varnish. Although it forms in many environments on earth, from arctic to humid, it forms best in hot deserts and therefore is often called desert varnish. Rock varnish, composed primarily of clay minerals colored dark brown to black by oxides of manganese and iron, forms most commonly on hard, silica-rich rocks. The longer these rocks are exposed on a desert land surface, the darker they become. Several hypotheses of varnish formation have been circulated for many years, but it is generally agreed that the constituents of the varnish are derived primarily, if not entirely, from sources external to the rock they cover, for even the whitest sandstones—whether they be massive cliffs or small pebbles on the ground—can have this thin, dark mineral coating. The various hypotheses state that: (1) The darker iron and manganese minerals may be present in the sandstone and collect on the surface during weathering. (2) The minerals may wash over the edge of cliffs or percolate down through the sandstone from a dark rock or soil layer above and be concentrated on the sandstone walls. (3) Thin films of windblown clay may adhere to the sandstone walls absorbing any moisture present and drawing the manganese minerals to them through capillary action. The clay minerals may help deposit the dark manganese oxide that then cements the clay to the rock surface. (4) The varnish is formed by the concentrating and fixing of eolian (airborne) dust by slow-growing manganese-oxidizing bacteria in environments that receive intermittent flows of water over the host-rock surface.

Rock varnished surfaces provided an ideal billboard for early American Indians to scratch out messages and artistic expressions, and these rock pictures (petroglyphs) have been left for us to see on darkened rock exposures throughout the southwest. In studying this ancient rock art, researchers have found that some petroglyphs have become revarnished in time, and new petroglyphs have been scratched over older ones. From these relationships, they have determined that it takes about 2,000 years for the darker coatings to form in a hot, arid environment.

Photo: Mark Vollmer

Ancient Puebloan rock art on a surface of Aztec Sandstone black with rock varnish. Petroglyph Canyon, Valley of Fire State Park.

interval	cumulative	
0.2	31.4	Pass under Union Pacific Railroad overpass.
3.5	34.9	Frenchman and Sunrise Mountains come into view at 11:00. Although difficult to clearly see, the Muddy Mountains are at 9:00.
		Las Vegas Valley lies directly ahead, located in a wide valley formed by extension during the Cenozoic Era (17 to 10 million years ago). The Sandstone Bluffs in the Red Rock Canyon area can be seen in the distance on the far side of valley.
		The high mountains to the right of the highway are in the southern part of the Las Vegas Range. Most of the range visible from here is within the 1.4 million-acre Desert National Wildlife Range. This land was set aside in 1936 to protect desert bighorn sheep habitat. The western part of the wildlife range (slightly over half of it) is included within the Nevada Test and Training Range, creating the interesting situation of a wildlife refuge that is also a military bombing range.
2.1	37.0	At about 3:00, in the low hills below the radio towers, is the Dike mining district. The district was discovered in 1916; the Lead King Mine, the only mine of note, produced small amounts of lead from shallow underground workings.
0.6	37.6	Nellis Air Force Base is at 11:00. This is one of the busiest Air Force training bases in the country, and you will usually see lots of air traffic as you pass by. Planes from here fly north to the 3-million-acre Nevada Test and Training Range for gunnery and bombing practice. Many of these training missions use live ammunition, so the planes you see taking off are likely to be armed as if on a wartime mission. If you were close enough to the runways, you might see markings of any number of foreign countries on the planes as numerous joint training exercises are conducted here. And if you are really lucky, you might see the Thunderbirds, the crack Air Force aerobatics team, practicing, as they are based at Nellis. On the other hand, it might be best to just concentrate on returning safely to Las Vegas—this stretch of I-15 is very busy and inattention to the road would be dangerous.
		End of Trip 3.

Photo: Kris Pizarro

Fruit heads of Apache plume (*Fallugia paradoxa*).

Photo: Jack Hursh

Desert evening primrose (*Oenothera deltiodes*).

Photo: Joe Tingley

Beavertail cactus (*Opuntia basilaris*).

Photo: Carol McKim

Photo: Jack Hursh

Land forms in the Aztec Sandstone, Valley of Fire State Park.

Photo: Carol McKim

Photo: Jack Hursh

TRIP 4: LAKE MEAD, HOOVER DAM, AND NELSON

This trip, which begins and ends in Henderson, visits the River Mountains, Las Vegas Wash, the scenic west and south sides of Lake Mead, Hoover Dam, and Boulder City. It also includes an optional trip down Eldorado Valley to the old mining town of Nelson in Eldorado Canyon, west of Lake Mohave. Total round-trip mileage, including the trip to Nelson, is about 105 miles.

The geology viewed on this trip provides an excellent introduction to the geologic history of the Lake Mead area over the past 15 million years. If you had traveled this same route in middle Miocene time (15 to 12 million years ago), you would have seen an area dominated by broad lava shields, stratovolcanoes, volcanic calderas, and wide intermontane basins. Later, between about 12 and 9 million years ago, the area was disrupted as the upper part of the crust was pulled apart (extended) and cut by numerous faults. The rocks were broken, tilted, and moved, so that today most of the ancient volcanic features are difficult for the casual observer to recognize. To bring the story up to date, beginning about 6 million years ago, cinder cones and basaltic lava flows formed at Fortification Hill. These are the youngest igneous rocks in the Lake Mead area.

HENDERSON

Photo: Nevada Historical Society

The Basic Magnesium complex at Henderson in the 1940s.

Henderson came into being as a direct result of a critical need for magnesium during World War II. The United States needed incendiary bombs for its war effort, and magnesium is a major component of the bombs. A deposit of magnesite (magnesium-bearing ore), was developed near Gabbs, some 250 miles to the northwest in the central part of the state. The process to extract magnesium from its ore required large amounts of electricity, and a site was selected at present-day Henderson for a recovery plant because of its proximity to a major source of electrical power—Hoover Dam. Basic Refractories of Cleveland, Ohio, won the contract to develop the bombs, McNeil Construction of Los Angeles built the plant (the facility was known as Basic Magnesium), and production of magnesium began in 1940. After the war, the state of Nevada acquired the site and named it Henderson, after Albert Scott Henderson, a lawyer, district attorney, judge, assemblyman, state senator, and Clark County pioneer who was instrumental in negotiating to have Basic's facility become the property of the state. Magnesium is no longer produced here, nor are bombs, but the old war-time plant site has evolved into a major industrial complex.

Four companies now share the Henderson industrial complex. Pioneer Companies, Inc. uses rock salt to produce chlorine, caustic soda, hydrochloric acid (which uses chlorine as a raw material), and bleach. Using chlorine from Pioneer and titanium ore from Australia and other sources, TIMET (Titanium Metal Corp.) makes high-quality titanium metal for use in the aircraft and space industries. A by-product of the electrolytic titanium refining process is magnesium chloride, a dust suppressant commonly used on haul roads at some of Nevada's mining operations. Kerr-McGee Chemical Corp. makes ammonium perchlorate used in rocket fuel, manganese dioxide for dry-cell batteries, and boron trichloride, an alloy used in tennis racquets and computers. Chemical Lime Co. makes hydrated lime (calcium hydroxide) from quicklime produced at their plant at Apex northeast of Las Vegas. The common facilities and property of the complex are owned and managed by subsidiaries of Basic Investments, which provide services such as water and high voltage power distribution, railroad facilities, and access roads to the four manufacturing companies.

interval	cumulative	
	0.0	Begin at the intersection of the Boulder Highway (S.R. 582) and Lake Mead Drive (S.R. 147) in Henderson. To get to Henderson from downtown Las Vegas, drive south on I-515 (U.S. 93/95) to Exit 61, Lake Mead Drive. Proceed east 1.8 miles to the intersection with the Boulder Highway. From McCarran International Airport, take the Las Vegas Beltway (I-215) to Lake Mead Drive in Henderson and then continue through the I-515 intersection to the Boulder Highway. Set your odometer to 0.0 at this intersection and head northeast on Lake Mead Drive. *(GPS 154)*
1.0	1.0	There are excellent views of Las Vegas Valley and the Spring Mountains at 9:00. Frenchman Mountain and Lava Butte are at 10:00 and 11:00, respectively. The geology of the Spring Mountains is described in Trip 2; Frenchman Mountain and Lava Butte are described in Trip 3.
0.7	1.7	Along this section, Lake Mead Drive is on pediment and alluvial fan deposits derived from the River Mountains to the east (pediments and alluvial fans are shown in the sketch on page 105).
0.3	2.0	Note the steeply tilted Paleozoic sedimentary rocks of Frenchman Mountain at 10:00. The large alluvial fan formed by sediments spilling from the wash to the south of Frenchman Mountain is very noticeable from here.
0.2	2.2	Las Vegas Wash flows through the low area to the north (left), where vegetation is riparian and therefore greener than elsewhere in the desert landscape.
0.5	2.7	Intersection, Lake Mead Parkway (S.R. 564) and Boulder Highway (S.R. 582). The white rock at 11:00, exposed in the cut now occupied by the large, white water storage tank is limestone of the Miocene Horse Spring Formation. The limestone was precipitated in shallow freshwater lakes that formed when existing drainage systems were blocked by volcanoes erupting to the southeast in the River Mountains. The dark-colored rock downslope to the immediate right of the water tank is a basaltic lava flow that overlies the limestone. There is a considerable outcrop area of basalt but, from here, it is hidden by the roofs of houses in the development *(GPS 155)*.
0.5	3.2	Calico Ridge Drive is on the left *(GPS 156)*.

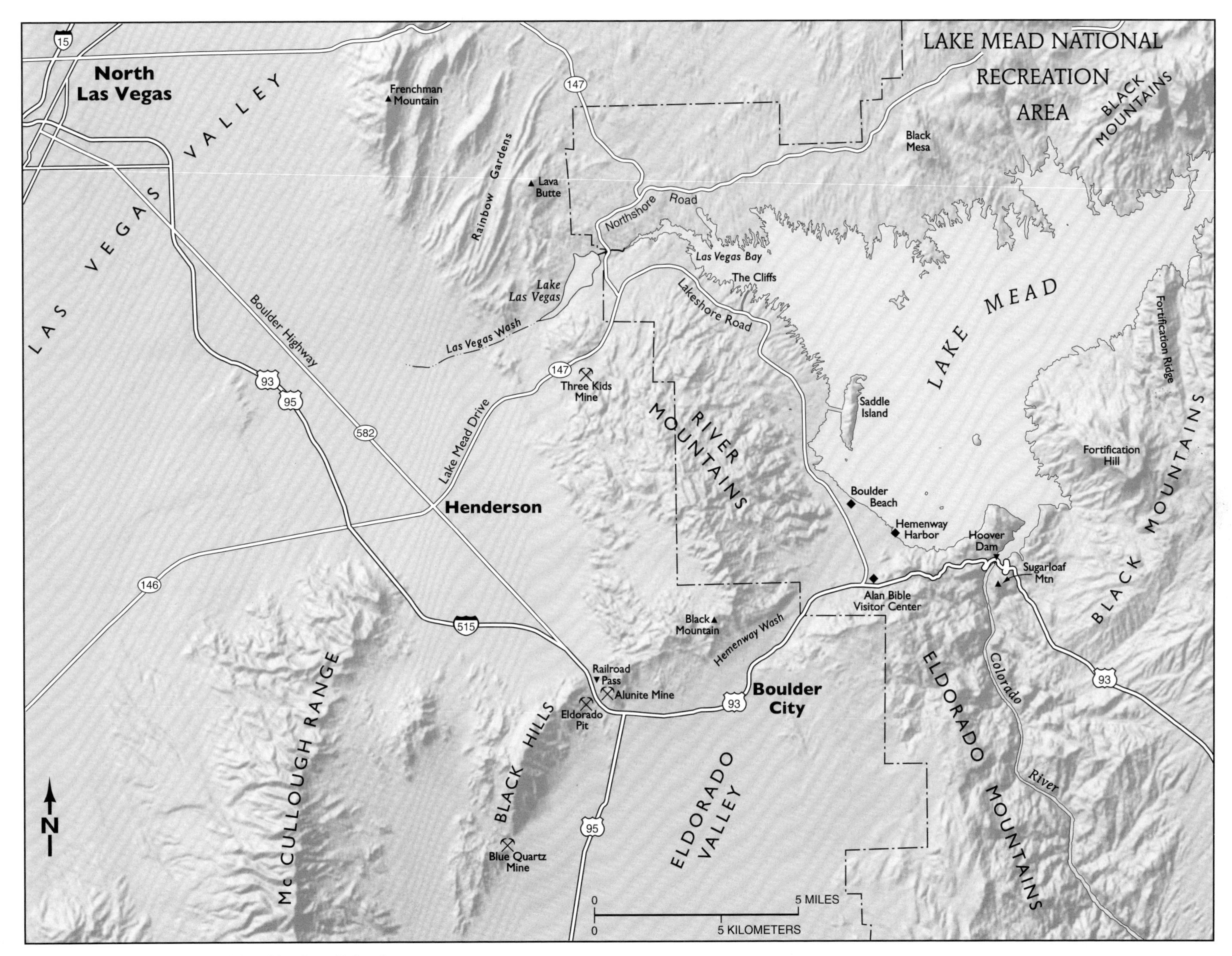

Route map, Trip 4 (see page 104 for side trip to Nelson).

interval	cumulative	
0.5	3.7	The River Mountains Aqueduct, the main water pipelines (there are two) of the Las Vegas Valley Water Project, pass beneath the road at this point (at Golda Way). The water intake is just southeast of Saddle Island in Lake Mead. Water is piped nearly 10 miles through the River Mountains to Las Vegas Valley. Approximately 70 to 80 percent of Las Vegas's water is delivered by this pipeline. The remaining 20 to 30 percent is pumped from deep wells in Las Vegas Valley. After the water is used, much of it is treated and returned to Lake Mead via Las Vegas Wash about 10 miles upstream from the Saddle Island intake area. *(GPS 157)*

The rocks on the skyline straight ahead are volcanic rocks of the River Mountains. Unlike most ranges that surround Las Vegas Valley, which are composed mainly of sedimentary rocks, the River Mountains contain mostly Tertiary volcanic rocks.

To the left in the valley beyond Las Vegas Wash is the colorful panorama of Rainbow Gardens. Most of the rocks visible from the road are tilted steeply to the east and are composed of sedimentary rocks deposited in broad basins 20 to 13 million years ago (Miocene time). Notice Red Needle Pinnacle at about 9:00. This "thumb" is an outcrop of the Thumb Member of the Horse Spring Formation and is capped by sedimentary breccia that is quite resistant to erosion. Large exotic blocks of rapakivi granite and gneiss are common in this unit. A lot of this rock formation is seen on Trip 3 where we describe it and its importance in the structural history of the area. Lava Butte is to the right of the "thumb" and the light-colored rocks between Lava Butte and the encroaching Lake Las Vegas development are limestone of the Horse Spring Formation.

interval	cumulative	
1.0	4.7	Lake Las Vegas Parkway on the left is the main access road to Las Vegas Wash and the Lake Las Vegas Resort. On the right are tailings deposits left by milling operations at the Three Kids manganese mine. *(GPS 158)*

Rocks to the north of the highway between this point and the mine entrance ahead are dacite flow rocks of the River Mountains.

interval	cumulative	
0.6	5.3	The entrance to the inactive Three Kids Mine and mill site is on the right. This is the largest of several manganese deposits located in the northern River Mountains. Old concrete structures are the remains of the milling plant. This is private property and is not open to the public. *(GPS 159)*

Photo: Becky Purkey

Water intake for Las Vegas's water supply at the southeast tip of Saddle Island. Black Mesa is in the middle distance and the Muddy Mountains are on the horizon (view is to the northeast).

Photo: Joe Tingley

View to the northwest of the River Mountains. Hemenway Wash crosses the photo from left to right in the middle foreground.

THE THREE KIDS MANGANESE MINE

In the late 18th and early 19th centuries, manganese, an important metal needed to strengthen steel, was supplied to United States manufacturers mostly by foreign mines. With the onset of World War I, these sources of manganese were cut off, and an intense search for domestic sources was undertaken. In Nevada, this prospecting effort led to the discovery of the Three Kids manganese deposit in 1917. This mine provided most of Nevada's manganese production during the last two years of WW I and operated intermittently until 1961. This is the largest manganese mine in Nevada and, during the 44 years of operation, more than 2,225,000 tons of ore ranging from 15 percent to 40 percent manganese was mined and treated here. Small amounts of lead, copper, silver and gold were also recovered by the mining operation.

The Three Kids deposit is a blanket-like layer (called stratiform by geologists) of earthy-appearing, brownish-black wad (an amorphous form of manganese oxide), with occasional streaks and grains of pyrolusite (black, crystalline manganese oxide). In places the wad is partially opalized (flooded with a type of silica) forming a hard glassy rock. Other manganese minerals found here are psilomelane, manganite, and neotocite.

The deposit is believed to have been formed by the flow of manganese-rich fluids into basins adjacent to zones of rapid uplift and detachment faulting, in this case a basin on the flank of the uplifted River Mountains. The fluids moved along faults cutting the basin and emerged at surface as hot springs which deposited the aprons or layers of manganese minerals. Volcanic ash erupting from volcanoes in the River Mountains at the same time then covered the manganese deposit. Minerals in the ash have been dated using the fission-track technique and suggest that volcanism and the formation of the manganese deposit took place between 12 and 14 million years ago. Other minerals mined from similar Tertiary basin sediments in the Lake Mead area include borates, gypsum, and salt (all described in Trip 3).

The Three Kids Mine is on private land and access is strictly controlled. This and any other abandoned mines you might come across can be extremely dangerous. Resist any temptation you may have to explore these workings—"Stay out and stay alive!"

Photo: Becky Purkey

Open pit at the Three Kids Mine. Note the folded sedimentary rocks of the Pliocene-Miocene Muddy Creek Formation forming a broad syncline exposed in the walls of the pit.

Photo: Becky Purkey

High-angle normal fault forms the wall of a large open pit at the Three Kids Mine. At this locality, manganese-rich sedimentary rocks of the Muddy Creek Formation (on the left) are faulted against Tertiary volcanic rocks.

interval	cumulative	
0.3	5.6	The route now crosses numerous lava (dacite) flows and sedimentary deposits related to volcanism in the River Mountains. Most of the low shrubs along the road here are creosote bush; they bloom with small yellow flowers in the spring and supply a very definite pungent odor to the air following every desert rainstorm.
0.5	6.1	State Historical Marker No. 141, Old Spanish Trail (Armijo's Route). *(GPS 160)*
0.1	6.2	Intersection with Lorin L. Williams Parkway to the left. S.R. 564 ends here. *(GPS 161)*
0.1	6.3	Enter Lake Mead National Recreation Area. Lake Mead Parkway (S.R. 147) now becomes Lakeshore Road. *(GPS 162)* There is an excellent view into Rainbow Gardens at 9:00.The PABCO Gypsum mine and mill are in the far distance (described on Trip 3, page 60). The Muddy Mountains are at 1:00. The Muddy Mountains contain thrust faults that place older Paleozoic limestone on top of younger Jurassic Aztec Sandstone. These faults are similar to (and are possibly separated segments of) those exposed in the Spring Mountains (refer to Trips 2 and 3 for descriptions of the faults in the Muddy Mountains and Spring Mountains. Lake Las Vegas and Las Vegas Wash are at 10:00.
0.1	6.5	Just before the paved turnout on the right, notice the low wire fence bordered by a line of evenly spaced rocks. This is a tortoise fence, built to prevent desert tortoises from wandering onto the highway. You will see these fences in many places along the road ahead. *(GPS 163)*

VOLCANOES IN THE RIVER MOUNTAINS

The River Mountains, to the right (south) of the highway, are composed of tilted and faulted mid-Miocene (15 to 12 million years old) andesite and dacite lava flows. Two volcanoes, the source of most of the lava, have been identified in the River Mountains. The first, located just north of Boulder City, is a stratovolcano surrounded by numerous domes. Stratovolcanoes are cone-shaped features composed of lava flows interbedded with abundant agglomerate and breccia that formed from explosive eruptions. The surrounding domes formed by the eruption of very sticky lava such as dacite, which erupted to the surface but did not flow. Lava formed a low spine or a dome that quickly crumbled to a ring of debris about the dome.

The second volcano, located in the northern part of the River Mountains, is a shield volcano—a broad volcano composed of basaltic and andesitic lavas similar to those erupting today on the island of Hawaii.

Both volcanoes have been tilted, cut by numerous faults, and broken into numerous parts producing a jigsaw pattern of rocks that is now very difficult to recognize.

Another type of volcano is the caldera. Calderas are associated with violent eruptions of ash, rock fragments, and pumice which produce ash-flow tuffs (commonly rhyolite or dacite in composition). Following eruption of this material, the empty magma chamber beneath the caldera collapses leaving a steep-walled basin-shaped depression at the site. Action may not end at this point, however. More magma may move into the chamber, forcing the collapsed rock in the basin to move upward like a huge plug. The resulting feature is called a resurgent caldera. There are no calderas in the River Mountains but there are in the Eldorado Mountains and the McCullough Range which we will pass at the end of this trip.

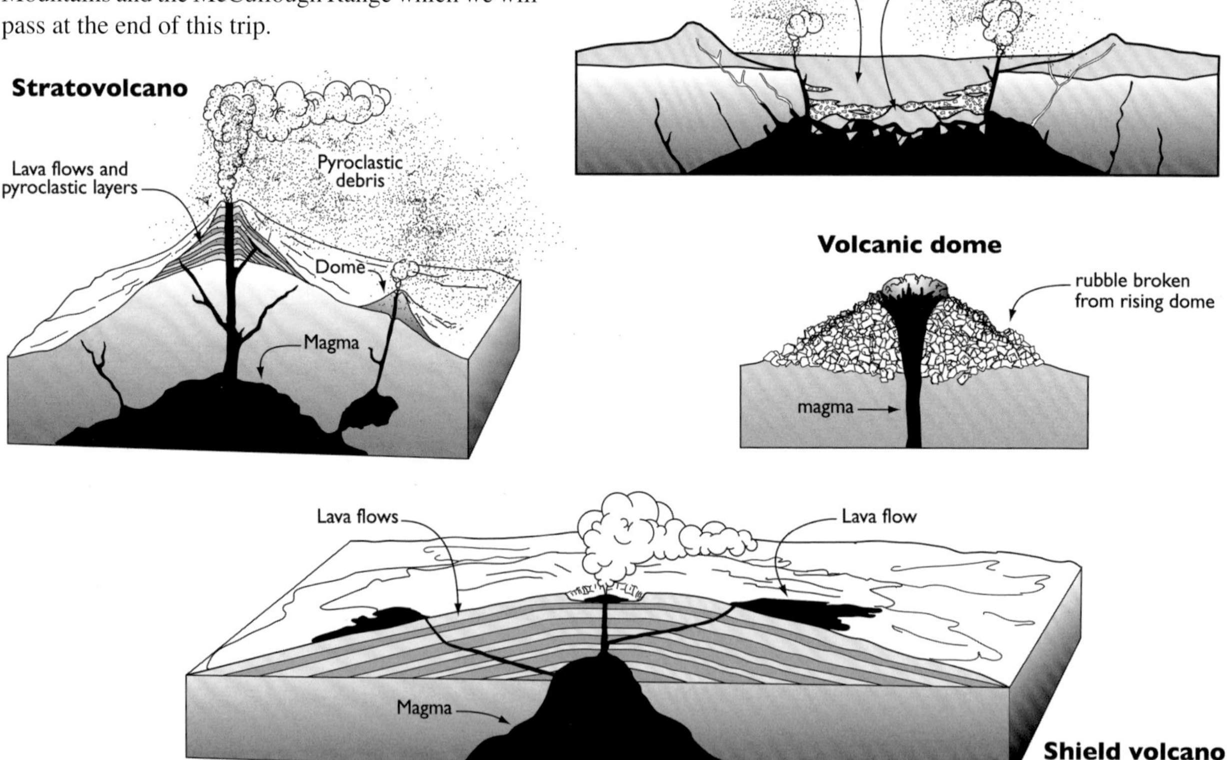

Desert Tortoise

The desert tortoise, Nevada's official state reptile, is the only naturally occurring tortoise in the Mojave Desert and is on the U.S. Fish and Wildlife Service's Endangered Species List. Over 6 million acres within the Mojave Desert, including over 1 million acres in southern Nevada, have been designated as critical habitat for the desert tortoise (areas that require special consideration and protection). This special consideration includes construction of tortoise fences in areas where highways or construction sites conflict with the tortoise lifestyle.

Desert tortoises are quiet and unassuming creatures, they spend most of their lives underground in winter dens and summer burrows where they are protected from the temperature extremes of the Mojave Desert. They are long-lived vegetarians, and are so efficient in their water usage that they can survive for more than a year without access to "free" water (water derived from sources other than the plants that they eat).

The desert tortoise population has been decreasing in recent years throughout the western Mojave Desert. One of the major factors in this decline is thought to be destruction of tortoise habitat by urban development. This is especially critical in areas surrounding Las Vegas where the city is rapidly expanding into prime tortoise habitat.

If you are lucky enough to see one of these elusive desert dwellers, don't touch it or pick it up (they tend to "lose" water if picked up—you might find this offensive but to the tortoise it could be fatal to lose even this form of stockpiled water).

Desert tortoise (*Gopherus agassizi*).

Photos: Susan Tingley

Tortoise fence.

interval | cumulative

0.2 | 6.7 Fee collection station for Lake Mead National Recreation Area. Stop and pay an entrance fee *(GPS 164).*

0.4 | 7.1 Junction of Lake Mead Drive with Northshore Road. To the left at 9:00 is the Lake Las Vegas Resort. *(GPS 165)*

SIDE TRIP TO LAS VEGAS WASH

0.0 Turn left on Northshore Road for a 2-mile round trip to Las Vegas Wash. *(GPS 166)*

The prominent hill ahead is Lava Butte (see Trip 3 for a detailed description of this feature). Lake Las Vegas is ahead on the left.

1.0 | 1.0 Cross the bridge over Las Vegas Wash and park on the left (west) side of the road. The flat bottom of this wash is characteristic of washes, gullies, and arroyos in arid regions (see page 63, mile 7.8). *(GPS 167)*

Photo: Becky Purkey

The bridge over Las Vegas Wash. Graffiti and elevation markings were painted on the bridge support column when protective rock riprap was still in place. Piled around the column as a protective measure, this material has since been washed away. There has been about 20 feet of down-cutting erosion by the stream since 1983.

LAS VEGAS WASH AND LAKE LAS VEGAS

Our view of the upscale Lake Las Vegas Resort complex, containing a 350-acre, 2-mile-long lake, and major hotels, golf courses, residential areas, and shopping centers, is far different than what was seen by Lt. Joseph C. Ives of the Corps of Topographical Engineers during his "Exploration of the River Colorado of the West." While attempting to navigate up the Colorado River, Lt. Ives reached Las Vegas Wash on March 12, 1858, believing it was the mouth of the Virgin River. In the account of his expedition, Ives noted "The appearance of the bed and the banks indicated the existence, during some seasons, of a wide and deep river. It was now but a few inches deep. The water was clear, and had a strong brackish taste. This fact, and its position, led me to suppose that we were at the mouth of the Virgen [sic], but I could scarcely believe that that river could ever present so insignificant an appearance." Then and there he decided not to ascend the Colorado any further, concluding his exploration of what he assumed was the navigable portion of the Colorado. With respect to the surroundings, and the view, Ives commented "Not a trace of vegetation could be discovered, but the glaring monotony of the rocks was somewhat relieved by grotesque and fanciful varieties of coloring." Las Vegas Wash was not the Virgin River, but Ives was correct when he observed that the wash carried a lot of water during some seasons.

Lower Las Vegas Wash drains the entire Las Vegas Valley and has done so since early Pleistocene time (about 1.8 million years ago). Looking upstream, the exposed sediments reveal a complex story of deposition and erosion extending back in time several million years. The oldest rocks exposed in the wash are red sandstone of late Miocene age (12 to 9 million years old).

The most conspicuous unit exposed in the wash is the coarse gravel named the conglomerate of Las Vegas Wash. It was deposited in a fast-moving stream with a steeper gradient than today as indicated by the size of the cobbles and the layering of the beds.

The cobbles are derived from the surrounding ranges and perhaps as far away as the Spring Mountains. They include all three major rock types: igneous rocks (granite, diorite, granodiorite, dacite, black fine-grained basalt and vesicular basalt with holes formed by escaping gas bubbles, and greenish-gray andesite), sedimentary rocks (limestone, dolomite, and sandstone), and metamorphic rocks (schist and gneiss).

Normally, this kind of young deposit can't be dated using current methods; however, researchers were able to determine the age of this formation by dating volcanic ash layers interbedded in the conglomerate. The distinctive chemical composition of the ash identified it as one that had erupted from Yellowstone Park 610,000 years ago. Thus, the conglomerate and the environment existing during its deposition could be dated. It is possible that at the time this conglomerate was being deposited high-energy streams flowed directly from the Spring Mountains into this wash. The ledges of conglomerate found in Kyle Canyon today (seen on Trip 2) may be the same conglomerate exposed here.

Photo: Joe Tingley

Las Vegas Wash upstream from the bridge on Northshore Road. The coarse gravel conglomerate of Las Vegas Wash overlies the light-colored, bedded, finer-grained sediments visible from water level to about half way up the exposed cliff face.

About 8,000 years ago, the southwestern United States experienced a great change in climate, and winter precipitation was drastically reduced. The wet climate of the late Pleistocene began changing to the arid climate that exists today. Along with this change in climate, perhaps between 4,000 and 1,000 years ago, Las Vegas Creek became an ephemeral stream, that is, it flowed only during storms when there was enough rain to fill the channel.

Since the late 1960s, the stream has been out of equilibrium. With every major flood, the stream deepens and lengthens its channel upstream toward Las Vegas. This is called headward erosion. The volume of sediment eroded from lower Las Vegas Wash from 1973 to 1988 exceeds the volume of concrete in Hoover Dam.

If left alone, the stream will eat its way into Las Vegas, adjusting its gradient so that flood waters will not move so fast as to erode more sediment than they deposit on their way into Lake Mead. However, in this process destruction of manmade structures and natural habitats could occur. Millions of dollars have already been spent to repair bridges and culverts and to protect pipelines.

Current developments in Las Vegas have caused Las Vegas Creek to flow constantly from treated wastewater discharged upstream. This large amount of flow, about 900 gallons per second, along with the increase in floodwater that is discharged into the wash during storms due to increased paving in the valley, has caused considerable downcutting of the wash. At this locality, erosion has lowered the elevation of the stream by about 20 feet since 1983. The summer floods of 1984 were responsible for about half of this total erosion, dumping several million cubic yards of sediment into Lake Mead.

Photo: Joe Tingley

View of the concrete mouth of the pipes that carry the water of Las Vegas Wash beneath Lake Las Vegas. Lake Las Vegas is impounded behind the dam seen in the background of the photo.

Treated wastewater (amounting to about three times the water taken from wells in the valley) flows down Las Vegas Wash into Lake Mead, which is the primary source of water for Las Vegas Valley. By federal agreement, Nevada receives an allocation of water from Lake Mead and is credited for water returned to the lake. That is, effluent that flows down Las Vegas Wash into Lake Mead allows Nevada to take more water from Lake Mead.

Lake Las Vegas now occupies the lower part of Las Vegas Wash that is adjacent to the Lake Las Vegas Resort complex. The resort lake is created by a 4,300-foot-long, 135-foot-high earthen dam and is filled with water pumped up from Lake Mead. Treated wastewater from Las Vegas passes beneath the lake through two 7-foot diameter pipes and empties into the wash downstream. In addition to the wastewater, these pipes or conduits are designed to adequately handle flood waters passing through Las Vegas Wash. The outlets for these conduits can be seen at the base of the dam.

Photo: Joe Tingley

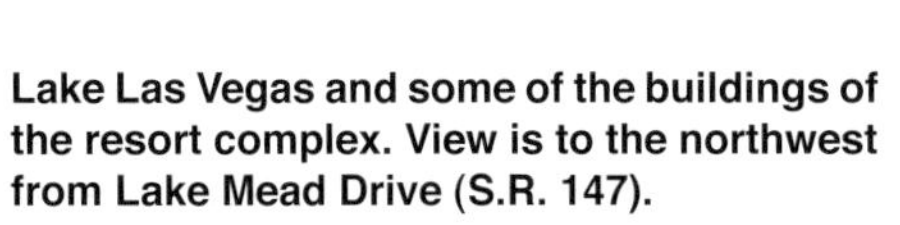

Lake Las Vegas and some of the buildings of the resort complex. View is to the northwest from Lake Mead Drive (S.R. 147).

interval	cumulative	
	0.0	Return to the intersection of Northshore Road and Lake Mead Drive. Reset odometer to 0.0. Turn left onto Lake Mead Drive and continue eastward. *(GPS 168)* Notice the tortoise fences on the right. Also note the desert pavement formed along both sides of the road. Dark, angular rock fragments weather out from underlying sediments—the softer, finer material blows away leaving the angular rock fragments behind appearing to have been set there by hand. But don't be fooled! Some of the new road cuts in this area have been "landscaped" by road crews who have set out rocks to look like desert paving. Its easy to spot the fakes, however, since nature rarely uses even spacing.
0.6	0.6	The route is now crossing sedimentary rocks of the Muddy Creek Formation. Muddy Creek sediments were deposited between 10 and 5 million years ago during late Miocene to early Pliocene time. This unit erodes easily, forming badlands topography and bluffs in the Las Vegas Wash–Lake Mead area. Note the scattered fragments of selenite (gypsum) that sparkle in the sunlight.
0.2	0.8	Look to the left, down the wash. You can see slightly tilted beds in the Muddy Creek Formation (tan and reddish rock) overlain by recent pediment gravels. There are more tortoise fences to the right. *(GPS 169)*
0.3	1.1	Lake Mead is straight ahead. Black Mesa is at 11:00. Fortification Hill, at 2:00 on the far side of Lake Mead, is capped by lava flows of basalt that are about 6 million years old. At 12:00, on the horizon, is Fortification Ridge in the northern Black Mountains in Arizona. Wilson Ridge, on the horizon to the right of Fortification Ridge, is composed of a Miocene (13.4-million-year-old) quartz monzonite pluton. A pluton is a body of once molten rock or magma that solidified below the Earth's surface. In this area many plutons represent the magma chambers that existed at shallow depths beneath the Earth's surface and supplied the molten lava for local volcanic eruptions. Notice the elongate, low, brownish hills about halfway up the slope beyond the trees that mark Las Vegas Bay. These hills are composed of fault-bounded slices of Paleozoic limestone that have been cut and shuffled along the Las Vegas Valley shear zone.

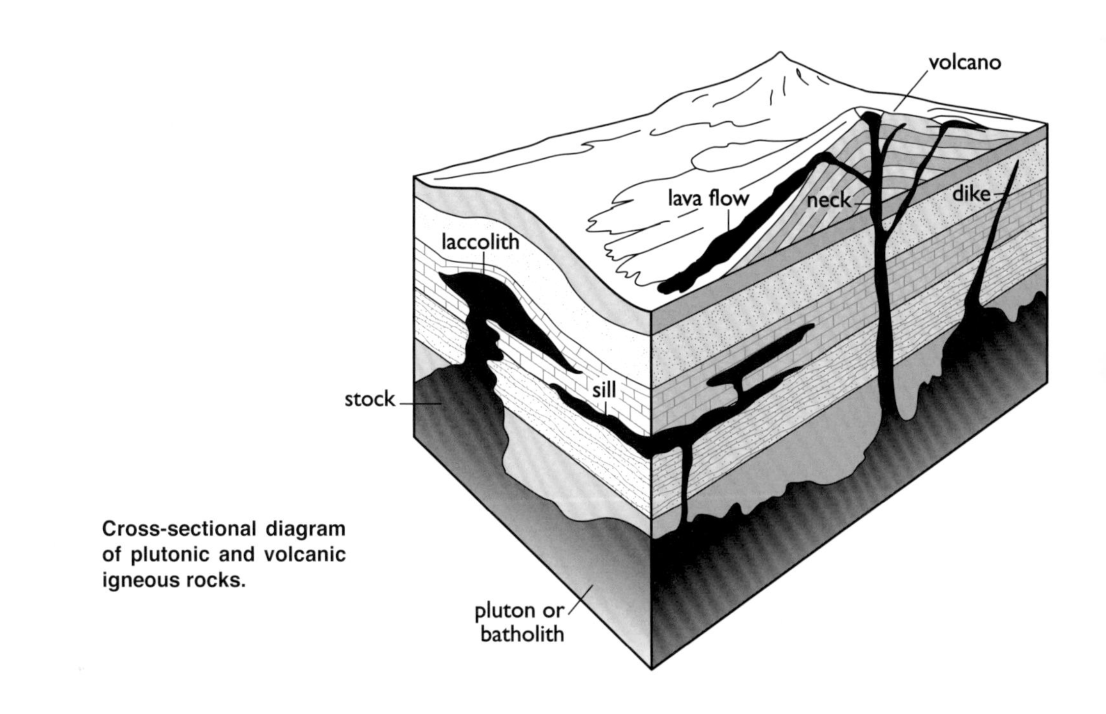

Cross-sectional diagram of plutonic and volcanic igneous rocks.

Aerial view east over Lake Mead. Fortification Hill (middle) is capped by flows of basalt. Light-colored granitic rocks of Wilson Ridge are in the background.

interval	cumulative	
0.7	1.8	Lake Mead Drive becomes Lakeshore Road at the intersection with the road (left) to the Las Vegas Bay Ranger Station, marina, and campground. *(GPS 170)*
0.2	2.0	For the next several miles the route crosses sheets of gravel shed from the eastern side of the River Mountains. These gravel sheets form a thin mantle over the volcanic bedrock and are called pediment gravels. In this area the pediment gravels cover older gravel and sand deposits of the Muddy Creek Formation.
		Look into the washes to your right as you pass through this section of road. Almost every wash has good exposures of Muddy Creek mantled by the younger gravel deposits. The Muddy Creek, here partly cemented, erodes to sheeted rugged cliffs, and the washes cutting through these sheets have formed spectacular, narrow canyons.
0.1	2.1	Intersection on the left with the road to scenic overlook of Las Vegas Bay Marina (0.3 mile to the paved parking area). This overlook comes equipped with covered picnic tables and rest rooms. *(GPS 171)*
1.4	3.5	The road to the left leads to the scenic overlook of The Cliffs (33 Hole Picnic Area). It's only 0.4 mile to three paved parking loops located close to the lakeshore, each with its own complement of informational signs, picnic tables, and restrooms. The Cliffs are formed where canyons cut in the resistant gravels of the Muddy Creek Formation have been drowned by the water of Lake Mead. There are vertical cliffs at the water's edge, and even small islands of Muddy Creek Formation (or peninsulas, depending on the lake level) extending from shore. Also of interest, the exposures of the Muddy Creek Formation here contain several volcanic ash layers. The age of these volcanic deposits is unknown, but the ash may have drifted to this area from large eruptions to the north in the vicinity of the Nevada Test Site. *(GPS 172)*
0.6	4.1	Paved road to the left leads to another scenic overlook of Lake Mead. *(GPS 173)*
		At 3:00 on the skyline are the faulted remains of the volcano from which much of the basalt in the northern part of the River Mountains was erupted. About 12 million years ago it may have resembled the broad shield volcanoes on the island of Hawaii.
0.9	5.0	This area contains several small volcanoes or volcanic domes of Miocene age. The dark red and brown rocks that you see exposed along the road on the left are the eroded remnants of these small domes and the dacite and basalt flows that came from them.
0.1	5.1	Desert View scenic overlook on the left. *(GPS 174)*

Photo: Becky Purkey

Pediment gravels shed from the eastern side of the River Mountains (poorly sorted gravel in the top third of the bank) cap sediments of the Muddy Creek Formation (the bedded rocks exposed in the lower two-thirds of the bank).

Photo: Becky Purkey

Eroded central crater area of the River Mountains shield volcano.

interval	cumulative	
0.8	5.9	Junction with road to the Lake Mead Fish Hatchery. *(GPS 175)*
0.1	6.0	The road to Saddle Island is to the left; at 10:00 (before you go into the road cut) there is a good view of the northern part of Saddle Island. This part of the island forms the upper plate of the Saddle Island detachment fault (sketch at right; see discussion of detachment faults on page 108). Notice the prominent saddle on the north side of the island. The detachment fault is exposed on the north side of the saddle and divides the island into two plates. Note the textural and color changes associated with the fault. The upper plate rocks, on the left of the saddle, form light red, jagged, rugged outcrops, and the lower plate rocks on the right (south) of the saddle form a smooth-appearing greenish outcrop. *(GPS 176)*

Recently (in 1987—recent in geologic time) geologists proposed that the volcanoes of the River Mountains were once positioned over the granites now exposed on Wilson Ridge (the prominent ridge across the lake at about 11:00). The Wilson Ridge granites, therefore, may represent the magma chamber, once far below the surface, that fed volcanic rocks now seen in the River Mountains. These geologists (Mike Weber and Eugene Smith of the University of Nevada, Las Vegas) believe that about 13.4 million years ago, the Saddle Island detachment fault formed, beheading the surface volcanoes from their deep source. The Wilson Ridge pluton and the volcanoes, lava flows and all, were carried in the upper plate of this fault about 12 miles westward to become the present River Mountains.

Access to Saddle Island is limited and there is a locked gate on the west side of the causeway leading to Saddle Island. Permission to walk the causeway to the detachment fault may be obtained from the offices of the Southern Nevada Water Authority.

interval	cumulative	
0.2	6.2	Straight ahead is a panorama of the northern part of the Eldorado Mountains. This part of the range is composed of the Boulder City pluton, a middle Miocene-age quartz monzonite intrusion. The dark, conical peaks in the first row of outcrops to the left of the highway are dacitic volcanic rocks of the Hoover Dam volcanic series. At 1:00 is the rugged core of the River Mountains stratovolcano.

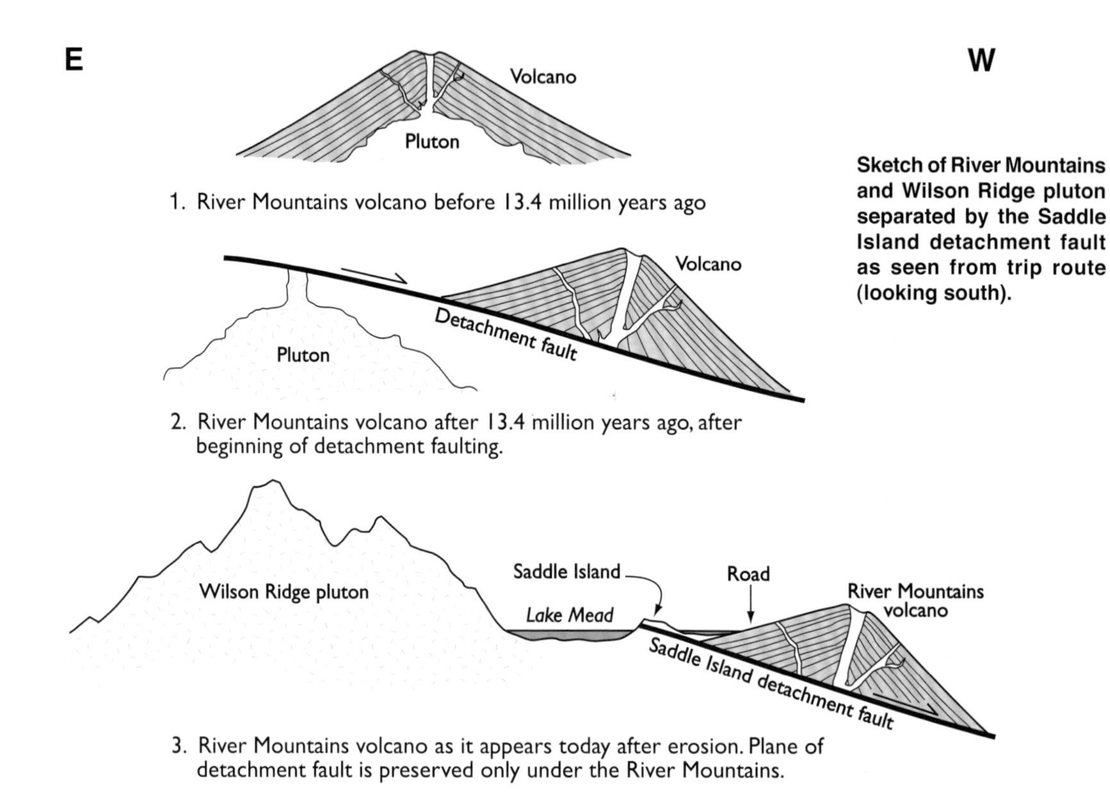

Sketch of River Mountains and Wilson Ridge pluton separated by the Saddle Island detachment fault as seen from trip route (looking south).

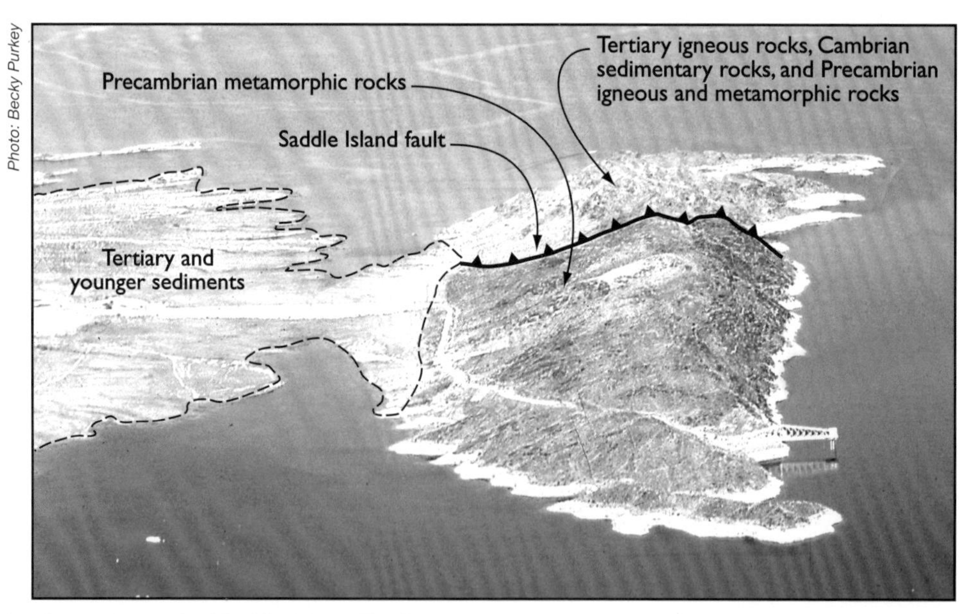

Aerial view to the north of Saddle Island. The Saddle Island detachment fault separates Precambrian metamorphic rocks in the lower plate on the south from a complex of younger rocks in the upper plate to the north.

interval	cumulative	
0.5	6.7	The road now leaves the rolling terrane characteristic of the Muddy Creek Formation and enters the relatively flat alluvial deposits derived from the River Mountains.
0.1	6.8	Entrance to the Alfred Merritt Smith Water Treatment Facility to the left. *(GPS 177)*
0.9	7.7	Note the quarry in the steep canyon at 3:00 (look for a fresh-appearing "flash" on the canyon face). The volcanic unit quarried here is the dacite of Teddy Bear Wash, named for a stand of teddybear cholla (cactus) in the eastern River Mountains. The quarried rock was used to construct the causeway to Saddle Island, and was also used as fill to reconstruct the southern breakwater at Lake Mead Marina.
0.5	8.2	Lake Mead Marina to the left. *(GPS 178)*
0.1	8.3	Entering the congested area near Boulder Beach and Lake Mead Marina. The road to the right goes to National Park Service employee dwellings.
0.4	8.7	Junction with road to Lake Mead Lodge on the left. Ahead to the right are two isolated buttes of dacite and dacite breccia. Volcanoes are relatively short-lived phenomena. They erode rather rapidly and commonly are surrounded by aprons of debris eroded from their flanks. The massive cliffs behind the two buttes represent the debris apron that at one time existed around the River Mountains stratovolcano. *(GPS 179)*
0.4	9.1	The road on the left is an entrance to Boulder Beach and picnic area. Between this point and about mileage 9.4 ahead, there is a great view of Fortification Hill to the left. Dark lava flows cap lighter tan and red sedimentary rock, and talus cones of the darker material can be seen spilling down the cliffs covering the lighter rocks. *(GPS 180)*
0.3	9.4	To the left is the road to the Boulder Beach Campground and Trailer Village. *(GPS 181)*
0.3	9.7	In the foreground at 3:00 is the eroded central crater area of the River Mountains stratovolcano. The light-colored rocks are the River Mountains quartz monzonite pluton that occupies the vent area of the volcano. The reddish rocks above the intrusion are altered andesite and plutonic rock cut by numerous dacite dikes that emanate outward from the River Mountains pluton. The alteration was caused by heat from the intruding pluton and by circulating warm water. The alteration products are primarily iron oxide and clay accompanied by barite, specular hematite (a dense, hard form of iron oxide that is commonly made into beads and other forms of jewelry), and manganese oxides. The dark rocks at the summit are andesite and dacite flows that formed on the flanks of the stratovolcano. This view of the River Mountains provides a excellent geologic section through the core of a volcano.
		At this point the road crosses the trace of the Hamblin Bay fault, one of the strike-slip faults that collectively form the Lake Mead fault zone. You cannot see this fault (it is buried under gravel) but it follows Hemenway Wash and may be responsible for the abrupt and steep truncation of the River Mountains north of the wash.

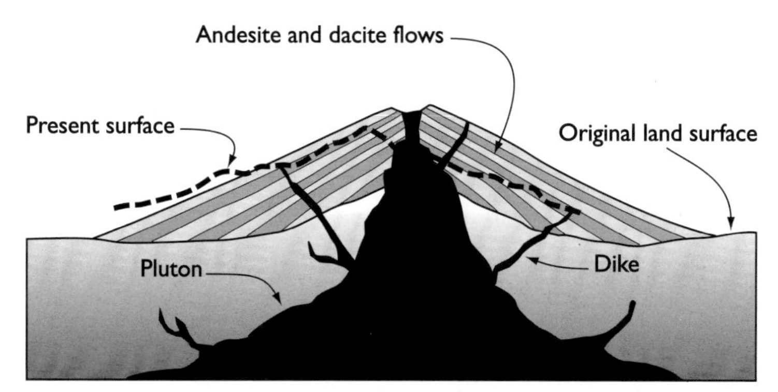

Generalized geologic section through the core of the River Mountains stratovolcano.

Photo: Becky Purkey

Looking west at the flanks of the River Mountains stratovolcano north of Boulder City.

interval	cumulative	
0.2	9.9	The route now traverses modern gravel deposits of Hemenway Wash. On the skyline at 2:00 note the water storage tank and homes in Boulder City.
		Straight ahead is a good view of the Boulder City pluton. The pluton varies in composition from diorite to quartz monzonite and has two topographic expressions. The uppermost exposures are composed of smooth rolling hills and the lower part is composed of rugged cliffs. Various ideas have been proposed to explain the differences in the two outcrop areas of the same rock. One offered by R.E. Anderson of the U.S. Geological Survey suggested that the difference in appearance of the rock is due to the presence of a paleohydrologic surface (an ancient water table). Rocks formerly below the ancient surface have been highly altered by circulating warm water while the rocks formerly above it are relatively fresh. This surface forms a "bathtub ring" along the south side of Hemenway Wash.
0.3	10.2	The road on the left is the access route to Hemenway Harbor. *(GPS 182)*
0.2	10.4	Fee station for the Lake Mead National Recreation Area is on the left. We now leave the Recreation Area. *(GPS 183)*
0.2	10.6	At 9:00 to 11:00 are exposures of highly altered dacite flows overlain by (or in fault contact with) dark-gray to black flows of dacite.
0.4	11.0	Trailhead parking on left for the Historic Railroad Trail. This trail follows a section of railway bed used during construction of Hoover Dam. The trail has excellent views of Lake Mead and its wide and nearly level—easy for both cyclists and hikers. The trailhead and parking are just down the hill from the Alan Bible Visitor Center. Toward Hoover Dam, a one-way trip of about 2.7 miles will take you through five tunnels and eventually the trail will continue down to the dam. In the other direction, you can also follow the trail 3.6 miles to Boulder City via the old railroad grade. *(GPS 184)*
0.1	11.1	Turn left into the Alan Bible Visitor Center for the Lake Mead National Recreation Area, or proceed ahead to the junction of Lakeshore Road with U.S. 93. *(GPS 185)*
		If you stop at the visitor center, take time to browse through the interesting displays about the human and natural history of the Lake Mead area. The geology exhibit is very well done and includes a geologic map of the Lake Mead area, a description of the various rock formations, and samples of key rock types. There is a desert garden, a well marked nature trail, and a great view of Saddle Island and Lake Mead from the observation deck. The bookstore at the visitor center has a complete selection of books and maps that describe the resources of the recreation area. Drinking water and restrooms are also available here.
0.3	11.4 0.0	Intersection of Lakeshore Road and U.S. 93. Reset your odometer to 0.0, turn left on U.S. 93 and proceed to Hoover Dam. *(GPS 186, 187)*
0.9	0.9	Hacienda Hotel and Casino is on the left and a gas station is on the right. *(GPS 188)*
		For the next mile, the road follows a strike-slip fault that lies within the Lake Mead fault zone. Intrusive rocks of the Boulder City pluton are faulted against Miocene volcanic flows and sedimentary units. Note the numerous fault surfaces with slickensides (smooth grooves in the fault surface) and dark red to purplish iron-oxide staining in exposures on the north (left) side of the highway.
		From here, the road begins to descend to Hoover Dam. The deep canyon on the right provides a popular route to several hot springs located just below Hoover Dam.

Photo: Claire Lanouette

The Alan Bible Visitor Center, Lake Mead National Recreation Area.

Photo: Becky Purkey

View to the east toward the Hacienda Hotel and Casino. Dark Miocene volcanic rocks form the rugged hills in the center of the photo. Light-colored granitic rocks of the Wilson Ridge pluton are on the distant horizon and part of Fortification Hill is in view on the left edge of the photo.

interval	cumulative	
0.3	1.2	Intersection, new U.S.93 Bypass on right (approximate location, under construction). *(GPS 189)*
0.1	1.3	Excellent view of intrusive rocks of the Boulder City pluton in the canyon on the right.
0.3	1.6	S-curves on U.S. 93. Note the polished and striated fault surfaces (slickensides) in exposures on left side of road. Be careful that you also watch the road, many accidents have occurred here!
0.2	1.8	There is a narrow paved turnout on the right where you can safely pull out and view the fault surfaces with slickensides mentioned above. Be very careful, especially if you decide to cross the highway for a closer look at the faults. Traffic here is fast and very heavy.
0.8	2.6	Turn off to Lakeview Overlook on the left. This stop provides a panorama of Lake Mead, Saddle Island, and Hemenway Wash. *(GPS 190)*
		Past the road to Lakeview Overlook, the nature of the rocks changes completely. The first exposures of a geologic unit called the tuff of Hoover Dam are encountered. This volcanic unit erupted explosively about 14.3 million years ago (during Miocene time) and may have originated from a caldera (a large volcano that collapsed after expelling its ash and volcanic flow rocks) just northwest of Hoover Dam.
0.1	2.7	Lower Portal Road to the right is a direct route to the base of Hoover Dam, but access is strictly controlled by the U.S. Bureau of Reclamation. On the left side of the road is the main warehouse for Hoover Dam. *(GPS 191)*
		For the next mile the road cuts through exposures of the tuff of Hoover Dam.
		It's said that this is a good area to watch for desert bighorn sheep at sunset, but don't look while you drive. The sheep are elusive and hard to spot.
0.2	2.9	Security checkpoint. *(GPS 192)*

Photo: Becky Purkey

HOOVER DAM

Hoover Dam is named after Herbert Hoover, 31st President of the United States. Construction on Hoover Dam began in 1931—during the Depression—and was completed in 1935, two years ahead of schedule and under budget! The many stories of leadership and engineering technology—from the planning through building stages of the dam—are well worth reading. The dam is 726 feet high, 1,244 feet long at its crest, and 660 feet thick at its base. The dam was built for flood control on the Colorado River and hydroelectric energy production, and it produced Lake Mead, America's largest manmade reservoir. The lake was named for Dr. Elwood Mead, Reclamation Commissioner during the dam's construction. The depressing side to this engineering wonder is that magnificent canyons full of beauty and archaeological and historical riches were removed forever from the enjoyment and wonder of future generations. Of the electricity produced here, only 8.8 percent is allotted to southern Nevada.

Tours of Hoover Dam are conducted daily; tickets may be purchased at the visitor center and exhibit building on the Nevada side.

Aerial view of Hoover Dam (left).
Hoover Dam construction in the mid-1930s (below).

Nevada Historical Society

Desert bighorn sheep *Artwork: Ralph Bennett*

interval	cumulative	
0.4	3.3	The road begins its steep descent to Hoover Dam. Note the excellent exposures of fault surfaces on the right side of road. Sugarloaf Mountain is at 12:00 across the Colorado River in Arizona. It is composed of dacite flows at the top and the tuff of Hoover Dam below.
0.7	4.0	On tight curve, a basalt dike cuts the tuff of Hoover Dam in exposures on the left side of highway.
0.2	4.2	Entrance to covered parking garage is on the left. If you prefer to park in Nevada, you may want to take advantage of this multilevel parking garage (there is a parking fee). From the parking structure, you can walk to the new visitor center where you will find exhibits relating to the history and construction of the dam and perhaps sign up for one of the tours that take you deep into the working heart of the structure. In addition, there is a bookstore and gift shop to check out. To keep your odometer on track with the trip log, however, you will still need to cross Hoover Dam, turn around in Arizona and return so you can reset your odometer at the described point (the winged bronze statue). *(GPS 193)*
0.1	4.3	Begin crossing Hoover Dam.
0.1	4.4	Entering Arizona. *(GPS 194)*
0.1	4.5	Just ahead, notice the well-exposed polished and striated fault surface. The road sign directional arrow must have been positioned by a geologist. The arrow corresponds to the direction of motion of the fault (right-lateral). *(GPS 195)*
		Beyond the dam, turn left into one of the parking lots. Park your vehicle and walk back to the dam.
	0.0	When you have finished touring and browsing, reclaim your car and drive back across the dam into Nevada. Reset your odometer to 0.0 at the winged statue on the north side of the dam and continue on to Boulder City. *(GPS 196)*
1.4	1.4	The lower portal road is to the left. On the right, in the large open area beside the highway, was the location of the plant that produced cement used in the construction of the dam. During World War II, this was the inspection area for vehicles crossing the dam. Security was high since there was the threat of sabotage of the dam and its vital output of electrical power. Cars and trucks were inspected here, then shuttled across the dam in convoys accompanied by armed military personnel—no stopping on the dam was allowed!
0.2	1.6	The Lakeview Overlook is on right.
1.6	3.2	The Hacienda Hotel and Casino is on the right, and a gas station is on the left.
0.4	3.6	From the turnout on the right, there is a good view of Lake Mead and Saddle Island at 3:00.
0.5	4.1	Junction of Lakeshore Road with U.S. 93. Continue straight ahead on U.S. 93.
0.2	4.3	At 2:00 there are good views of the eroded central crater area of the River Mountains stratovolcano.

Worker's tool check badge used during construction of Hoover Dam.

The face of Hoover Dam and the power plant below.

Photos: Mark Vollmer

Hoover Dam and Lake Mead at night, taken from the Arizona side. The construction crane on the Nevada side in the background marks the site of the now-completed visitor center.

interval	cumulative	
1.2	5.5	"Welcome to Boulder City" sign. *(GPS 197)*
0.2	5.7	Boulder City, city limit sign. *(GPS 198)*
0.2	5.9	Pacifica Way, River Mountains Trailhead parking to right. *(GPS 199)*
0.6	6.5	Intersection of business route to Boulder City on the left and U.S. 93 bypass straight ahead Continue straight ahead. *(GPS 200)*
0.7	7.2	The turnoff to the River Mountains Hiking Trail is on the right where there is a sign along with a small parking area and display. The hiking trail follows trails blazed by the Civilian Conservation Corps (CCC) in the 1930s and continues for nearly 3 miles to the summit of Black Mountain. There are excellent views of Lake Mead from the summit. The trail ascends through Red-Black Canyon to the saddle just below the Boulder City VORTAC, an aircraft navigation facility at the summit of what is locally known as of Red Mountain (neither Red-Black Canyon nor Red Mountain are official names, see comment on trip 1, page 21). A major fault in the canyon places Miocene volcanic rocks of the River Mountains stratovolcano (black colored) against highly altered older Miocene volcanic rocks. Paleozoic limestone and dolomite of the Mississippian Callville Formation locally crop out on the east side of Red-Black Canyon and are the remnants of a mountain range that once existed in this area. The range was buried by volcanic rock and intruded by rising magma during the formation of the River Mountains stratovolcano. Follow the trail to the east through the saddle and climb through flows on the flanks of the River Mountains stratovolcano. The hike is recommended only for individuals in good condition. You will need sturdy hiking shoes or boots and plenty of water. The best time to hike this trail is in winter, November through April, because mid-day temperatures during the summer may reach 110°F. *(GPS 201)*
0.7	7.9	Intersection of the U.S. 93 bypass with the Boulder City business route. Turn right at the freeway interchange onto U.S. 93 to Las Vegas. *(GPS 202)*

BOULDER CITY

Boulder City was originally built to serve as a supply center for construction of Hoover Dam and to house the workers who flooded here to find work during the Depression. Some workers were lucky enough to live in the dormitories provided by the Six Companies, Inc., the consortium of companies who joined forces to construct the dam.

At the end of 1935, as the dam neared completion, the workers began to leave, looking for new work. Buildings were dismantled or moved to other parts of the state; the trees and grass died. In some areas only paved streets and sidewalks remained. But the U.S. Bureau of Reclamation kept their offices here to oversee the operation of the dam, and the U.S. Bureau of Mines opened a branch office for a few years to sample ores in the region. A few original residents stayed on, using materials from abandoned buildings to build their homes and churches. Boulder City remained a federal reservation (no gambling, no liquor) until 1960, when it became a regular chartered Nevada town. Facilities still remain limited, and gambling is still not allowed, but it is a charming town with a quiet and stable atmosphere. Property is much sought after by retirees wishing to live in a small-town environment.

Photo: Becky Purkey

Boulder City in 1993. View is to the southeast from the summit of Red Mountain in the River Mountains north of the city.

Nevada Historical Society

Panoramic view (to the south) of Boulder City, October 1933.

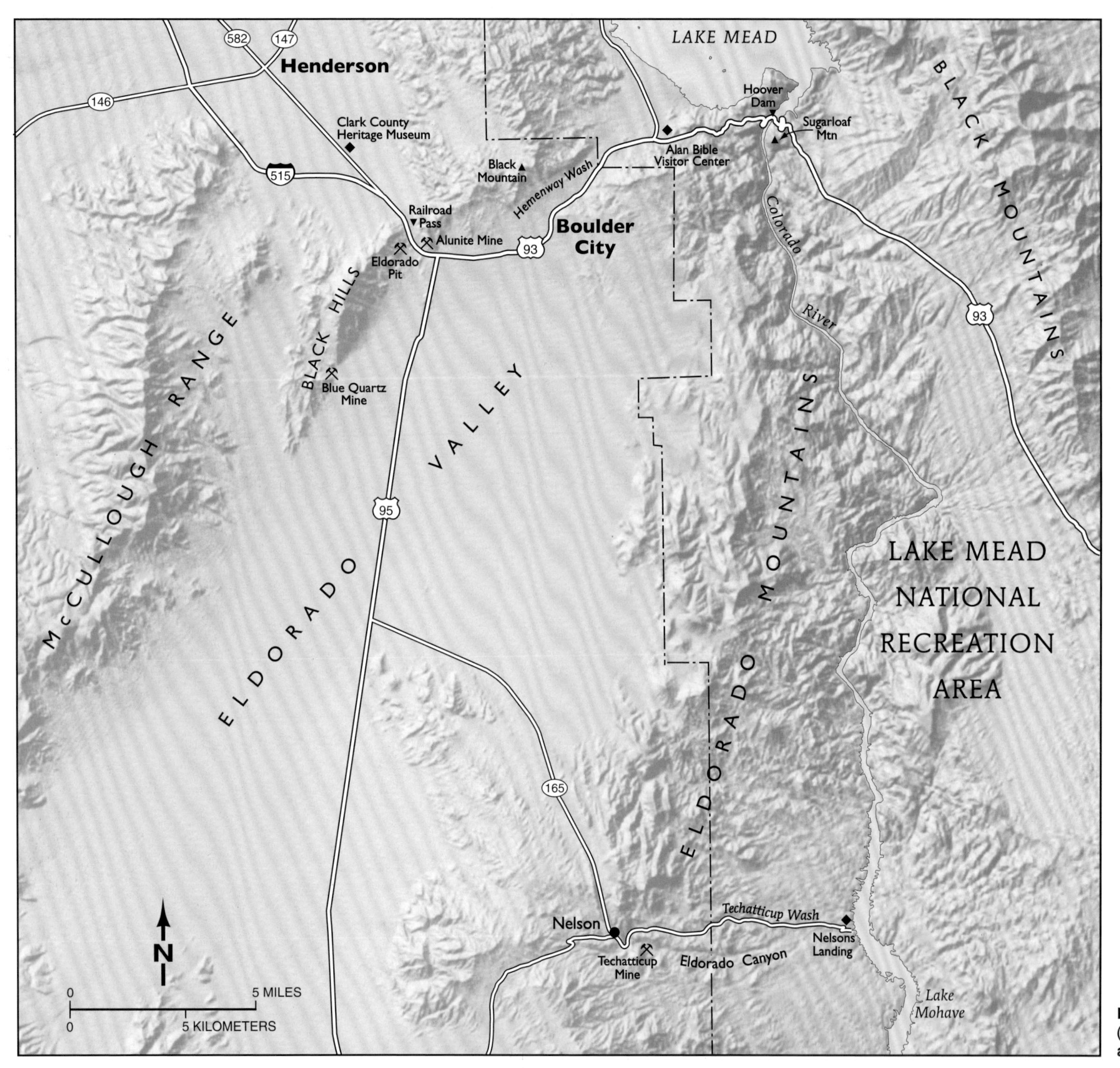

Route map, Trip 4. (side trip to Nelson and Eldorado Canyon)

interval	cumulative	
0.6	8.5	On the right, Yucca Drive provides access to the southern part of the River Mountains, the Bootleg Canyon Trails and the Nevada State Railroad Museum. *(GPS 203)*
0.5	9.0	The rocks to the right (north) of U.S. 93 are basalt, andesite, and dacite flows that dip gently to the east.

The jagged peaks of the Railroad Pass pluton are straight ahead. The pluton is probably part of the Boulder City pluton and is composed of granite and quartz monzonite.

1.8 | 10.8 Junction with U.S. 95. Exit to the right and turn left (south) at the bottom of the off ramp onto U.S. 95 toward Searchlight for the optional side trip to Nelson and Eldorado Canyon. Set odometer to 0.0. *(GPS 204)*

If you don't want to take the side trip now, continue ahead on U.S. 95 to Henderson (see page 113 for road log).

SIDE TRIP TO NELSON AND ELDORADO CANYON

0.0 The 58-mile round trip to Nelson begins by crossing the expanse of Eldorado Valley, a typical basin of the Basin and Range province with its north-trending mountain ranges separated by broad valleys. In this province, major faults generally form the borders of the mountain ranges. We will see the trace of one of these faults on the west side of Eldorado Valley. *(GPS 205)*

The McCullough Range, in the background on the west side of Eldorado Valley, is formed largely from volcanic rocks that originated from Miocene shield and stratovolcanoes (sketches of types of volcanoes are on page 92). The trip route then crosses through the Eldorado Mountains allowing us to see evidence of detachment faulting where Miocene volcanic rocks have slid as jumbled fault blocks from a Precambrian core. After passing through the historical Eldorado mining district, we follow Eldorado Canyon to the site of Nelsons Landing on Lake Mohave and become acquainted with some hoodoos along the way.

0.5 | 0.5 Ahead, the north-trending Eldorado Valley is typical of basins in the Basin and Range province in that it has no drainage outlet. All drainage is toward a lake bed, or playa, in the center of the basin. Alluvial fans spread from the mountain ranges toward the playa. Eroded bedrock mantled by alluvium close to the mountain front is called a pediment surface. A bajada is formed by coalescing alluvial fans. This valley is fault-controlled, that is, a fault runs down the long axis of the valley. It may be a southern extension of the Lake Mead fault zone.

This straight segment of U.S. 95 is constructed on the surface of a broad bajada that extends from the River Mountains into the Eldorado Valley. The Eldorado Mountains are to the east (left) and the Wilson Ridge is beyond, in the far distance in Arizona. The Highland Range is straight ahead and the McCullough Range is to the west (right). The steep rugged outcrops in the mountain just to the west (right) are granite of the Railroad Pass pluton.

Photo: Becky Purkey

Eldorado Valley playa (foreground) with the McCullough Range in the background to the west. Note the layered appearance of the volcanic rocks near the top of the range.

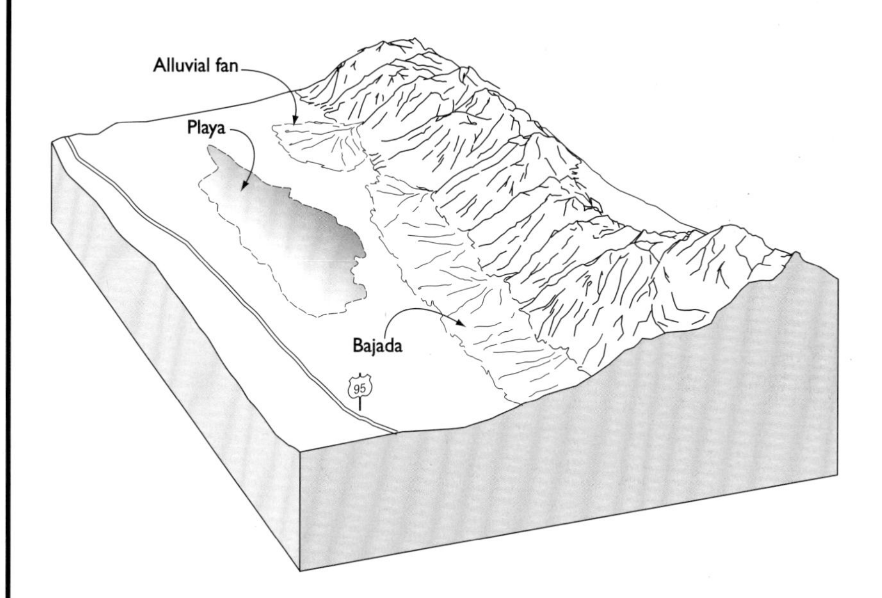

Landforms typical of the Basin and Range province.

Photo: Roy W. Cazier

With excellent vision and a wingspan of 6 to 7 feet, the **golden eagle** (*Aquila chrysaetos*) can often be seen soaring over open country. It can dive at speeds up to 200 miles per hour in pursuit of prey. It feeds on a wide variety of mammals, birds, reptiles, and carrion. Golden eagles are thought to mate for life. A pair often has several nests in an area and may alternate from year to year. Nests can grow as large as 10 feet in diameter and 3 to 4 feet deep.

Photo: Larry Anderson

Late afternoon view of the eastern side of the Black Hills on the west side of Eldorado Valley southeast of Las Vegas. The dark line that crosses the photo near the base of the range marks the shadowed offset of the ground along the Black Hills fault. This offset (or fault scarp) is about 7 feet high and is a relic of a magnitude ~7 earthquake that occurred between 5,500 and 8,200 years ago.

The **Mojave rattlesnake** (*Crotalus scutulatus*) grows to a length of 24 to 51 inches. It is generally greenish gray with white scales outlining brown diamonds along its back and alternating black and white rings around its tail. It is commonly found in open country in association with mesquite and creosote bush. Its venom is extremely toxic.

Photo: Red Rock Canyon National Conservation Area

interval | cumulative

1.5 | 2.0 Notice that the steep mountain range on the right (the McCullough Range) is truncated along a sharp line that forms the boundary between the range and the adjacent bajada. As explained above in the discussion on the Basin and Range province, this topography is the result of a steep fault at the range boundary. Look closely and, if the light is right (best in evening shadow), you may be able to see a linear fault line crossing the alluvial fan at the base of the mountain range, proof that faulting is still active here. Look for a dark, straight line that crosses all of the alluvial fans and small canyons. This is the trace of the Black Hills fault. The line you see is the fault scarp, a relic of an earthquake with a magnitude of about 7 that occurred between 5,500 and 8,200 years ago. *(GPS 206)*

1.9 | 3.9 Crossing under the high-voltage power lines that carry electricity from Hoover Dam to the McCullough switching station and then to Phoenix and Los Angeles. As mentioned earlier, Las Vegas receives little of its power from Hoover Dam. *(GPS 207)*

0.6 | 4.5 Bands on the mountain to right are the roots of a Miocene stratovolcano. The dark bands are dikes. There is also gold mineralization in the mountain (at the Blue Quartz Mine located on the mountain front at about 4:00). On the skyline (right) in the McCullough Range gently tilted lavas may represent the flanks of this volcano. The oldest volcanic flows are about 16 million years old and the youngest about 13 million years old.

1.4 | 5.9 The Eldorado Valley playa (on the right) is dry during most of the year and, when dry, is a favorite place to fly model airplanes and ultralight aircraft. However, after a major storm the playa may fill with several inches of water to become a large, shallow lake. The lake has no outlet and the water quickly evaporates in this arid climate, leaving its dissolved salts on the dry lake bed. This is the cycle described in Trip 3 (at mile 4.0) that may form gypsum and other deposits. Ephemeral lakes like this one are the habitat of the long-tailed apus, a shrimp-like crustacean that hatches from microscopic eggs when the lake fills with water. Adults mature quickly—reaching a length of about 2 inches—and lay their eggs before the lake completely dries again. The adults then die and the eggs are scattered by the wind. The eggs withstand months of dry heat before they again receive rain and the cycle from hatching to maturity to eggs repeats.

3.1 | 9.0 Volcanoes are especially well exposed in the McCullough Range to the west (right) because there has been little tectonic faulting and tilting. Volcanoes exist in the Eldorado Mountains to the east but have been dissected by numerous faults and tilted (in some areas to nearly 90°). Imagine trying to recognize a volcano that has been tilted on its side!

To the west, you can see the sequence of Miocene rock layers in the McCullough Range. The dark band at the range summit is andesite of the Hidden Valley volcanics; the white stripe is the McCullough Pass tuff erupted from a caldera in the McCullough Range and below are the Eldorado Valley volcanics. The McCullough Pass tuff (the white band) does not extend to the north. Its flow was stopped by a topographic barrier formed by a large stratovolcano in the northern McCullough Range.

The high southern section of the McCullough Range is composed of Precambrian rock (1.7 billion years old).

Long-tailed apus.

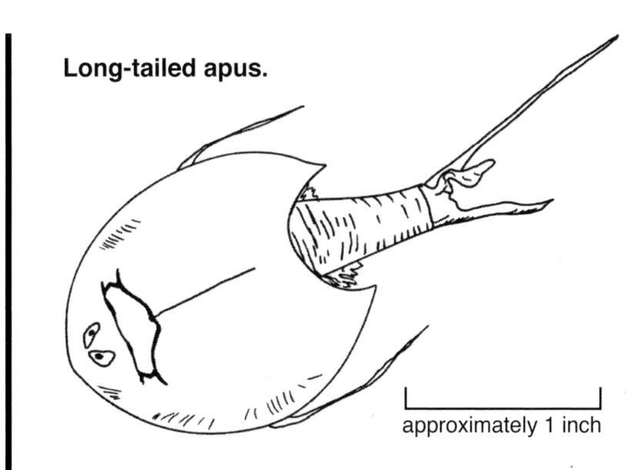

Cross sections of the Miocene McCullough stratovolcano, looking south. The upper diagram shows the stratovolcano in Miocene time, prior to erosion. The lower diagram shows the stratovolcano as it appears today, after 13 to 16 million years of erosion.

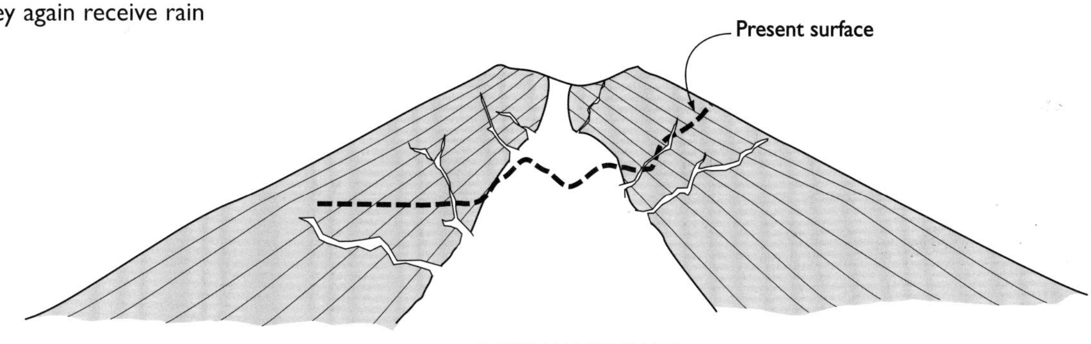

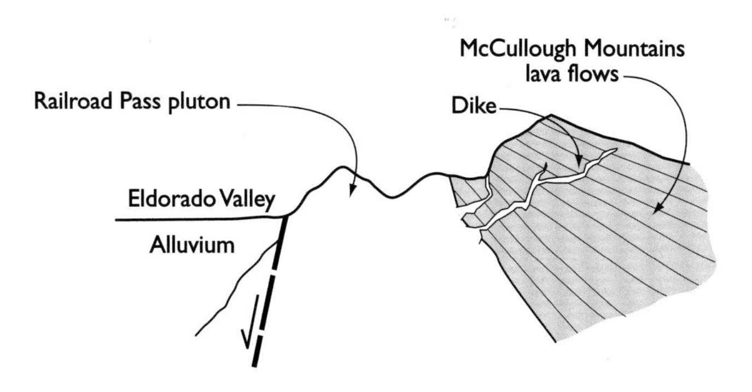

interval / cumulative

1.0 | 10.0 Intersection of U.S. 95 with S.R. 165. Turn left (east) onto S.R. 165 to Nelson. *(GPS 208)*

Note the historical marker (NHM #6) at the intersection describing the history of Eldorado Canyon and the Eldorado mining district. Written records show that mining began in this area in 1857 and has continued intermittently to the present. The total value of mineral production, mostly gold and silver, is $7 to 10 million. Mines in the Eldorado district have colorful names like Techatticup, Wall Street, and Honest Miner. The district is located near the contact of a large pluton with steeply tilted Miocene volcanic rocks. The gold and silver are found in quartz and calcite veins that cut shattered plutonic rock.

0.3 | 10.3 The road ascends a broad pediment surface that extends from the base of the Eldorado Mountains (ahead and extending to the left) into the north-trending Eldorado Valley.

The small hill to the left is called Bump and is composed of Miocene andesite. In this area, andesite flows are tilted steeply (65 degrees to 90 degrees) to the east and are cut by near horizontal faults.

At 9:00 are the River Mountains, Fortification Hill, and Wilson Ridge (on the skyline).

2.6 | 12.9 The road crosses steeply tilted exposures of andesite lava flows.

0.5 | 13.4 The powerlines overhead carry electricity from Hoover Dam to southern California. *(GPS 209)*

0.6 | 14.0 More andesite flows, steeply tilted to the east.

3.5 | 17.5 The small quarry cut into the side of the hill to the left exposes the tuff of Bridge Spring, a regionally extensive ash-flow sheet that erupted from a caldera in the Eldorado Mountains about 15.2 million years ago. The hills extending away from the road to the left are thought to be composed of a thick section of this tuff that was deposited against the wall of the caldera. The caldera wall, and the tuff, has been tilted 90 degrees onto its side. The tuff exposures are cut by a left-lateral strike-slip fault (such as faults of the Lake Mead fault zone seen in Trip 3) that represents an adjustment or tear fault in the upper plate of a major detachment sheet. This unit is named after Bridge Spring, located about 2 miles north of Nelson and about 1 mile east (left) of S.R. 165. *(GPS 210)*

DETACHMENT FAULTS IN THE ELDORADO MOUNTAINS

The extension of the crust during Miocene time (17 to 10 million years before present) was originally thought to be accommodated by steep normal faults. G.K. Gilbert, a geologist with the U.S. Geological Survey, proposed this idea in the late 1880s. In 1971, R.E. Anderson, based on work in the Eldorado Mountains near Nelson, suggested that crustal thinning was related to the formation of low-angle normal faults. Low-angle normal faults were recognized earlier (1960) by Chester Longwell, but Anderson's careful and insightful analysis of the rocks in the Eldorado Mountains was critical to demonstrating their importance during crustal extension. Low-angle normal faults like these that move rocks a considerable distance are called detachment faults. In the Eldorado Mountains, and in other areas in southern Nevada and western Arizona, detachment faults place younger rocks (commonly Tertiary volcanic rocks) in the upper plate over older rocks (usually Precambrian rocks) of the lower plate. These faults are sometimes called denudation faults because, when viewed in cross section, they give the appearance of blocks of younger rocks sliding away from a upward-bulging center of Precambrian rocks. As is the case in science, there are some geologists who disagree about the importance and even the existence of these faults.

Both the Eldorado Mountains and the McCullough Range lie in the upper plate of a detachment fault. The fault may be close to the surface in the Eldorado Mountains, causing surface rocks to fault and tilt, and may be deep beneath the McCullough Range to the west. As a result, the McCullough Range is being carried along in the upper plate of the fault as an intact block, and is neither strongly tilted nor faulted.

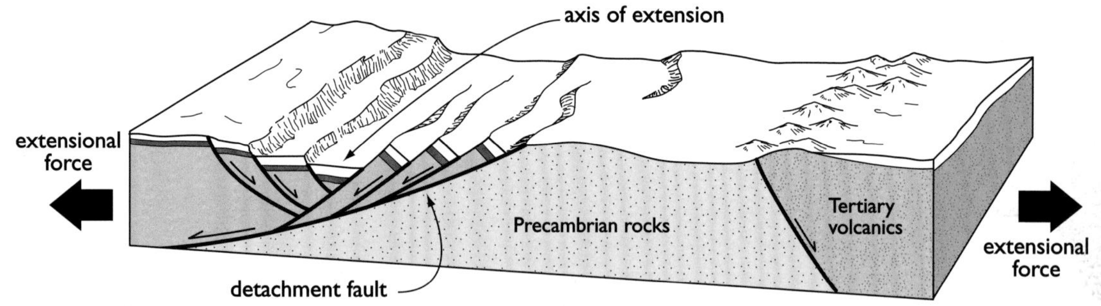

Block diagram of a detachment fault (low-angle normal fault).

interval	cumulative	
0.4	17.9	The tuff of Bridge Spring forms the jagged cliffs on the skyline to the east (left). The tuff dips about 45 degrees to the east and is underlain by basalt and andesite flows that were erupted between 18.5 and 15.2 million years ago.
0.5	18.4	The road crosses east-dipping reddish flows of basalt and andesite and interbedded lenses of white sandstone composed of rhyolite detritus.
0.5	18.9	The ridge on the right is composed of rhyolite domes and flows. Since the rock layers on the ridge are steeply tilted, there are excellent cross sections of volcanic vents displayed on the ridge. Unfortunately, these can only be observed from the air. The rhyolite contains abundant spherulites (spherical masses of radiating feldspar crystals, quartz, and jasper) which weather from the rock and accumulate in the washes below. Some of the spherulites are up to 3 inches across, and may have potential for cutting and polishing.
0.3	19.2	To the east (left) a dacite dike cuts across exposures of basalt and andesite flows and the tuff of Bridge Spring.
0.5	19.7	The route begins descending into Nelson. Steeply tilted lavas and tuff of Bridge Spring are on the left. Exposures of rhyolite are to the right.
0.8	20.5	Entering Nelson. Nelson is the center of the Eldorado mining district, one of the oldest mining districts in Nevada. S.R. 165 continues to the east (left) and begins a long descent through Eldorado Canyon which includes Techatticup Wash and several other washes in its drainage. Volcanic rocks are to the left and the Nelson pluton is on the right. *(GPS 211)*

Photo: Becky Purkey

Steeply dipping volcanic rocks in the Eldorado Mountains just north of Nelson.

ELDORADO MINING DISTRICT

Local legend has it that gold was discovered in Eldorado Canyon some time in the 18th century by the Spanish exploring the Colorado River, but the Mohave Indians were hostile, the Spanish moved on, and mining never progressed. The term "El Dorado," here shortened to one word—Eldorado, means "the gilded one" in Spanish, and was used by miners throughout the Americas to show faith in the richness of their mines.

Arrastres and prospect holes supposedly dating from the Spanish activity were found later, around 1857, when gold was rediscovered on the Honest Miner claim. Known today as the Eldorado mining district, the district was organized as the Colorado mining district in 1861 while the area was still part of New Mexico Territory.

Most of the district's total production in gold, silver, copper, lead, and zinc was mined between 1864 and 1900 from veins in quartz monzonite. The Techatticup Mine, the largest producer in the district, was responsible for more than $3,500,000 in production between 1860 and 1900, and an additional $1,117,200 from 1900 to 1930. The name techatticup may come from the Southern Paiute words tecahenga (hungry) and tosoup (flour) combined to mean "hungry, come and eat some flour." It may also mean "enough for everybody," "plenty for all," or "white flower."

To service the mines, the townsite of Eldorado grew on the Colorado River at the mouth of Eldorado Canyon. A 15-stamp mill was erected there in 1863 to treat ore from the Techatticup Mine. Salt needed for the milling process was mined from deposits on the Virgin River near St. Thomas and transported to Eldorado by boat down the Virgin and Colorado Rivers. In 1905, the townsite of Nelson was founded and old Eldorado was abandoned. The site of Eldorado is now beneath the water of Lake Mohave.

Many mine shafts, adits, and drifts honeycomb the hills surrounding Nelson. The old workings are not always protected by fences, so if you go exploring in these hills be careful to stay out of these old holes.

Nevada Historical Society

Operations of the Southwestern Mining Company on the shore of the Colorado River at the mouth of Eldorado Canyon in about 1890.

interval	cumulative	
2.1	22.6	Teddybear cholla cactus "forest" on both sides of the road (see color photo on page 119). Named for their innocent overstuffed appearance, with small arms and a dark bottom, these inhabitants of the lower Mojave and Sonoran deserts are not at all friendly if you mingle too closely with them. Like other members of the cholla family, the short segments of the arms break off and attach themselves to clothing and flesh if you brush against them. The thorns have short, almost invisible filament-like ends that will grab, giving the impression the cholla has "jumped" onto you. Hence the other name for this plant, "jumping cactus."
0.4	23.0	The Techatticup Mine is on the right. Discovered in 1861, the mine is responsible for over half of the total production of the Eldorado district. A cyanide mill operated on the property until 1942, but it has been idle since then. Most of the approximately 3 miles of underground workings are along the Techatticup and Savage veins, which crop out near the top of the rugged hills to the south of the old mill ruins at the mouth of the canyon on the right. If you look closely, you can see old pits, tunnels, and stopes as they trace the vein outcrop across the hills. *(GPS 212)*
		The old mine and building remains are private property, but if the idea of looking around underground appeals to you, check with the folks at the small store on the left. They offer guided tours (for a fee) of some of the old mine workings.
0.5	23.5	Look straight ahead at the ridge on the near skyline. You see the dark, jagged openings of an abandoned mine. These mark where a stope (the cavity left underground after miners have taken out the ore) has broken through or "daylighted" to surface. To the right, just below the highest point in the ridge, is the waste dump from another part of the mine. *(GPS 213)*
0.3	23.8	Look to the right for another view of the open stope. You can see the vein here, it trends east-west (about parallel to the ridge) and dips gently to the south. Remember, these and all old mine workings are very dangerous—keep out of them!
0.2	24.0	We now cross the Eldorado fault, passing from volcanic rocks (to the west, or behind us) into Precambrian metamorphic rocks to the east (ahead of us). The Precambrian rocks are fairly easy to distinguish because they are greenish and weather to form rounded, subdued outcrops, and because here they are fairly uniformly dotted with teddy bear cholla. The volcanic rocks form rugged brown outcrops with irregular weathering cavities and shed huge blocks into the wash below their outcrop. The Eldorado fault may be the master structure (detachment fault, see figure on page 108) in the Eldorado Mountains. The high-angle normal and strike-slip faults seen north of Nelson lie in the hanging wall (upper plate) of this fault. *(GPS 214)*
		Driving ahead through the lower plate Precambrian rocks, notice the darker dike rocks that cut the sheared gneiss and granite.

Photo: Joe Tingley

Teddybear cholla along the road to Nelson.

From a distance the **teddybear cholla** (*Opuntia bigelovii*) resembles its namesake. A closer look reveals that its jointed stems are densely covered with backward-hooked spines ½ to 1 inch long. At the slightest encounter the stems detach and embed themselves in the passerby. The spines provide both protection and a means of propagation for the cholla. When the offending stems are eventually dislodged and fall to the ground, they easily root and produce a new plant.

interval	cumulative	
0.4	24.4	Entering Lake Mead National Recreation Area. Notice the white plastic posts in the cholla grove up hill on the right (and outside of the Recreation Area boundary). These are claim posts marking the boundaries of mining claims. You may have spotted others like these around Nelson. *(GPS 215)*
0.2	24.6	Just ahead, about where the road curves to the left, we cross another fault contact and pass back into Tertiary volcanic rocks. We are now again in upper plate rocks of the detachment fault.
0.1	24.7	Notice the large blocks of volcanic rock in the wash to the right. They have fallen from the high cliff face above. Large blocks such as these, when moved downstream by floodwaters and deposited in the lower parts of the wash, become the caps for the hoodoos you will see ahead.
0.8	25.5	Lake Mohave appears ahead. The Colorado River widens here due to the damming of its waters about 45 miles to the south by Davis Dam at Laughlin.
0.5	26.0	Conglomerates representing debris shed in alluvial fans from the Eldorado Mountains are banked against Precambrian bedrock to the south (right) of the road. For perhaps tens of thousands of years during the wetter Pleistocene epoch (1.8 million to 10,000 years before present), boulders, gravel, and sand were deposited here as muddy debris flows. Note the beautifully developed erosional spires (known as "hoodoos") in the gravels of Techatticup Wash (see color photo 10b on page 118). Similar landforms of similar age are seen in Kyle Canyon on Trip 2. *(GPS 216)*
1.3	27.3	Excellent views of gravel deposits in Techatticup Wash. Fire Mountain is to the east across Lake Mohave. The Fire Mountain area may be part of a major volcano, the source for many lava flows in the region.

HOODOOS, PRECARIOUS ROCKS, AND SEISMIC RISK

The erosional spires that rise from the bottom of Techatticup Wash, formed by hard protective boulders capping pillars of less weather resistant gravel, are known to geomorphologists and geologists as hoodoos. Hoodoo, meaning a column, pinnacle, or pillar of rock, is derived from an African term signifying a fancied resemblance to animals and embodied evil spirits. Seismologists, being more down to earth, simply call these and similar features precariously balanced rocks—and they also have found use for them beyond their role as decorative scenery. Groups of precariously balanced rocks evolve naturally and stay balanced until they erode and wash away—unless they are shaken down by earthquakes. Since, in desert climates such as ours, it can take thousands of years for these features to erode, their mere presence is direct evidence that no strong earthquakes have occurred here since their formation.

Precariously balanced rocks, like the hoodoos we see here, therefore provide important information about seismic risk. If hoodoos are prominent, there hasn't been a strong earthquake in a long time, probably not in thousands of years. Areas where hoodoos could or should form, but there aren't any, probably have been hit by fairly recent earthquake activity.

Hoodoos in lower Techatticup Wash west of Nelsons Landing.

Photos: Joe Tingley

interval	cumulative
1.5	28.8

A paved road descends into Techatticup Wash to the former site of the small recreational settlement of Nelsons Landing Marina and Resort. *(GPS 217)*

Flash floods and accompanying mud and debris flows are common in desert regions, and this settlement was destroyed by such an occurrence in October 1974. Nine people died. The storm that brought the rain was an unusual cloudburst that originated over Eldorado Valley and moved southeast to Nelson. In one hour it provided half the annual precipitation for this area. Waters from Eagle and Techatticup Washes joined those of Eldorado Canyon in a narrow channel only a few thousand feet west of Nelsons Landing.

The floodwater reached a maximum depth of 20 to 30 feet at the narrowest point in the channel (about 100 feet wide) and carried mostly gravel-sized debris and a few boulders up to 4 feet in diameter. This rock debris formed deposits at least 12 feet thick at the mouth of the canyon. Most of the town was carried into Lake Mohave.

A geologist reading the rocks along the sides of the canyon, where there is cemented debris from many past floods, would have known that this would be a prime location for such an event. Indeed, a year prior to this flood, the National Park Service had warned the inhabitants of Nelsons Landing of the danger of living in this narrow canyon.

▲ **Nelsons Landing was located at the mouth of Eldorado Canyon. At that point, the cemented alluvial fan gravels in the canyon walls formed a narrow slot through which the entire runoff from a 22-square-mile drainage area flashed, demolishing the small settlement. The view is to the northwest, and the Eldorado Mountains are in the background on the horizon. *Photo courtesy of the California Division of Mines and Geology.***

▲ **Damaged mobile homes, which were parked near the bottom of Eldorado Canyon when the flood swept through, were a total loss. In all, about 31 homes were damaged or carried out into Lake Mohave and lost. *Photo courtesy of the California Division of Mines and Geology.***

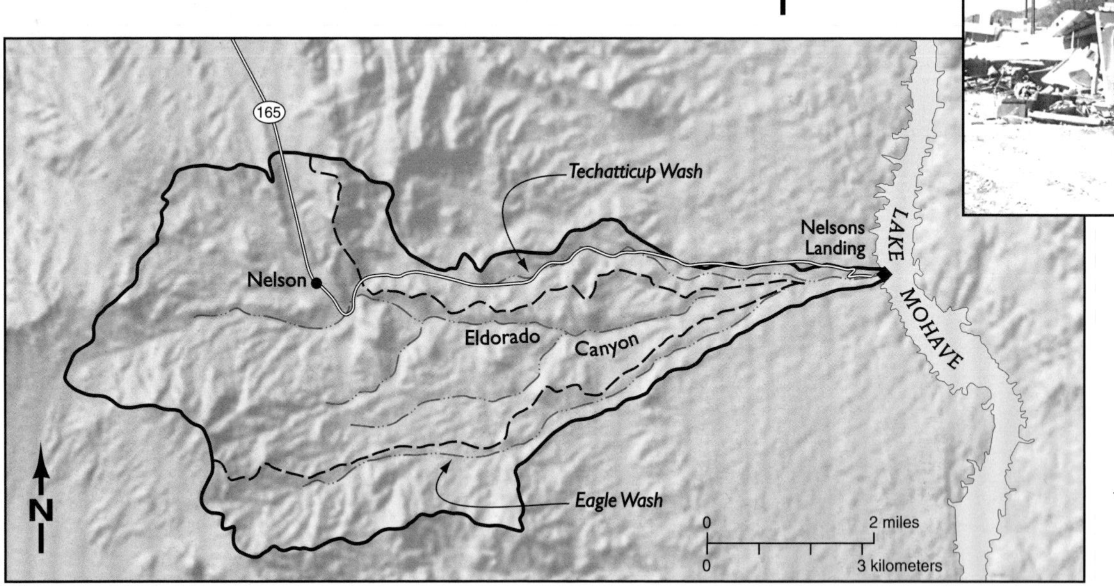

◀ **The Eldorado drainage basin is composed of three washes: Techatticup Wash on the north, Eldorado Canyon, and Eagle Wash on the south. These three canyons merge into one major channel about one mile west of Nelsons Landing.**

interval	cumulative	
0.2	29.0	Loop at the end of S.R. 165. Stop and view Lake Mohave. This area is a popular fishing spot and several trails lead down to the lake from here. Use caution if you decide to investigate them. *(GPS 218)*

This is the end of the side trip to Nelson. Retrace S.R. 165 back to the intersection with U.S. 95. Turn right on U.S. 95 toward Henderson and Boulder City.

CONTINUATION OF TRIP 4 TO HENDERSON

interval	cumulative	
	0.0	At the intersection of U.S. 95 and the Boulder Highway (U.S. 93), turn left toward Henderson and set odometer to 0.0. *(GPS 219)*
0.5	0.5	Large open-pit mining operation is visible at 10:00. This is the Eldorado Pit where granitic rocks of the Railroad Pass pluton are being mined for construction aggregate.
0.4	0.9	The Alunite (or Railroad Pass) mining district begins here and extends along both sides of the road. Small amounts of gold associated with alunite, a potassium-aluminum-rich sulfate mineral, were discovered here in 1908. Alunite was known to occur with the rich ores of Goldfield, to the north in Esmeralda County, and prospectors hoped the Alunite district would prove to be another Goldfield. Their hopes of a bonanza quickly faded. Later, during World War I, the alunite deposits were considered as a source of potash, a war-critical material and, during World War II, they again received attention as a potential source of critical aluminum. Neither flurry of activity resulted in any mining because the gold and alunite deposits are too small to be of value. You can, however, still see traces of the small adits and pits dug during the prospecting activity.
0.2	1.1	The Railroad Pass Hotel and Casino is on the right. *(GPS 220)*

The cliff above the casino is composed of andesitic lava and ash-flow tuff. The location of the volcano that erupted these units is a mystery because this volcanic section is separated from all other volcanic units by faults. An important northwest trending normal fault is exposed just below the casino water tank. This fault separates mineralized rock (west/left, in the Alunite mining district) from unmineralized rock (east/right).

There were many old underground mine workings in this area. Some penetrated the mountain as far as 500 feet and were very dangerous. In 1999 and 2000, the Nevada Division of Minerals, with the cooperation of the Bureau of Land Management and financial support from member companies of the Nevada Mining Association, backfilled and secured these workings.

State law now requires that open shafts be marked and fenced, but it is prudent to stay away from open shafts and take care when exploring in mining districts.

Photo: Becky Purkey

At Nelsons Landing, well-cemented blocks of alluvial fan gravel that caved off the canyon walls have covered and preserved gravel and boulder debris of past floods. Subsequent flood waters have banked up younger debris against these blocks on their upstream side (right of photo). Much geologic evidence here and in other parts of Eldorado Canyon suggests a long history of flooding. *Photo courtesy of the California Division of Mines and Geology.*

Photo: Joe Tingley

Mining for construction aggregate is underway at the Eldorado Pit.

interval	cumulative	
0.4	1.5	Railroad crossing. The tracks are paved over where the highway crosses them, but you can see them to the right, as the railroad grade passes above the casino. This railroad served as the main supply route to Hoover Dam during its construction. The tracks are no longer used, but plans to run a tourist train from Boulder City to Henderson through Railroad Pass have been proposed. *(GPS 221)*
0.8	2.3	Crossing Railroad Pass (elevation 2,373 feet). Straight ahead is an impressive view of Las Vegas Valley, especially spectacular at night. Beyond Las Vegas to the north lie the Sheep and Las Vegas Ranges, which form the northern boundary of the Las Vegas Valley.

The northern tip of the McCullough Range, beyond the TV towers at 11:00, is a caldera filled with dacite lava flows and domes (see figure on page 92 for sketch of a caldera). Named the Henderson caldera, it is about 6 miles in diameter and much of the new development in Henderson is built around the northern rim of the feature. The rounded hills behind the new subdivisions just south of Horizon Ridge Parkway are the lobate fronts of dacite lava flows originating from this caldera. The black-appearing mountain under the towers also is composed of dacite lava flows that flowed over the caldera rim into deep channels cut into older volcanic basement rocks. This caldera and its related volcanic flows are late Miocene age, probably between 15 to 12 million years old.

The northern McCullough Range contains many petroglyph sites, mostly in hard-to-find locations in isolated canyons. The early inhabitants may have spent the summer months in the mountains to escape the heat of Las Vegas Valley, but they were still close enough to take advantage of abundant water at the springs that until recently flowed in the valley.

The River Mountains are to the right of the highway (east). We found out all about these volcanic mountains earlier on this trip (refresh your memory by checking the description on page 92).

interval	cumulative	
0.8	3.1	Our route now becomes I-515. Take Exit 56B, Boulder Highway (S.R. 582). *(GPS 222)*
1.2	4.3	The Clark County Heritage Museum is on the right. Take time to stop. This museum offers realistic exhibits of the natural and historical highlights of this area. The displays change to highlight various topics and the bookstore carries a top line of informational books and gifts relating to this part of Nevada. *(GPS 223)*
2.9	7.2	Intersection of Lake Mead Drive and Boulder Highway in Henderson. End of Trip 4. *(GPS 224)*

Photo: Charles Webber, California Academy of Sciences

The **western chuckwalla** (*Sauromalus obesus obesus*) is a large, heavy-bodied lizard that inhabits rocky areas. It does not drink but relies on its diet of fruits, leaves, buds, and flowers for all its water intake. When alarmed, it wedges itself into rock crevices by inflating its body.

Photo: Joe Tingley

Clark County Heritage Museum.

Photo: Mark Vollmer

Lake Mead with the Muddy Mountains in the background.

Photo: Joe Tingley

View of Las Vegas Bay Marina from The Cliffs. Lava Butte is to the northwest in the middle distance, and Frenchman Mountain is in the background.

Photo: Becky Purkey

Black Canyon below Hoover Dam.

Photo: Becky Purkey

Aerial view of Hoover Dam and Black Canyon.

Color Photo Captions

PLATE 9

9a **Mouse's Tank**, Valley of Fire State Park. *Photo: Mark Vollmer*

9b **Black-collared lizard** (*Crotaphytus insularis*). *Photo: Jack Hursh*

9c **Elephant Rock**, Valley of Fire State Park. *Photo: Mark Vollmer*

9d **Brittlebush** (*Encillia farinose*) Common in the Sonoran and Mojave deserts, this member of the sunflower family prefers dry slopes and washes where it displays bright yellow flowers in mid to late spring. It grows 2 to 5 feet high forming a rounded, leafy bush with brittle, woody branches and long, slightly oval silver-gray leaves. The branches, which contain a fragrant resin, were chewed by American Indians and also burned as incense in the early mission churches of Baja California. *Photo: Patti Gray; inset: Jack Hursh*

9e Peaceful spring-fed pool below **Rogers Spring**, Lake Mead National Recreation Area. *Photo: Becky Purkey*

PLATE 10

10a **Rainbow Gardens**, **Lake Mead**, and the Black Mountains of Arizona viewed from the top of **Frenchman Mountain**. The area known as Rainbow Gardens is in the foreground and extends out of the photo to the right. **Lava Butte** is beyond, in right center. **Las Vegas Wash** angles across from the right, behind Lava Butte, and enters Lake Mead at Las Vegas Bay. Directly across Lake Mead is the distinctive black top of **Fortification Hill**, and Fortification Ridge forms the skyline to the left and behind Fortification Hill. *Photo: Mark Vollmer*

10b A **hoodoo** capped with a large boulder of volcanic rock in Techatticup Wash east of Nelson. Evidence of a wet spring is seen in the blooming brittlebush (in the center of the wash) and creosote bush (along the sides of the wash). *Photo: Joe Tingley*

10c **Red Needle Pinnacle** south of Lava Butte (in the background). The pinnacle is capped by a resistant conglomerate, called the Thumb breccia, that occurs within the Thumb Member of the Miocene Horse Spring Formation. *Photo: Steve Castor*

10d **Creosote bush** (*Larrea tridentata*) This large, evergreen shrub prefers well-drained plains and slopes and you will find it along the sides rather than in washes. It is the most characteristic species of the Mojave and Sonoran deserts. The shrub usually grows to less than 4 feet, but in favorable settings, can grow up to 12 feet. Creosote has yellowish-green, pointed leaves and small, yellow flowers, blooming from spring into summer. The resinous leaves have a pungent odor when crushed or wet. Like that of sagebrush of the northern Nevada deserts, the fragrance of rain-dampened creosote bush is one of the pleasant memories created by this low desert landscape. *Photo: Joe Tingley; inset: Jack Hursh*

PLATE 9

10a

10b

10c

10d

PLATE 10

11a

11b

11c

11d

11e

PLATE 11

12a
12b
12c
12d
12e
12f
N
Eglington fault
Windsor Park fault
Cashman Field fault
Valley View fault
Whitney Mesa fault
Subsidence (feet)
0
.3
.6
mi 3
0 km 5

PLATE 12

COLOR PHOTO CAPTIONS

PLATE 11

11a Along one of the paths on the grounds of the **Alan Bible Visitor Center**, Lake Mead National Recreation Area. Lake Mead is in the background. *Photo: Mark Vollmer*

11b A solitary **barrel cactus** (*Ferocactus acanthodes*) flanked by brittlebush on a sunny, south-facing slope of Paleozoic limestone near Rogers Spring in the eastern Muddy Mountains. The genus name, Ferocactus, comes from the Latin ferox ("fierce"), a name you will think is very appropriate if you accidentaly stumble into one. Barrel cactus store water in their pulpy flesh to get them through long dry spells, and after a wet season or a good rain their accordion-pleated skin expands to provide storage space. *Photo: Joe Tingley*

11c A stand of **teddybear cholla** (*Opuntia bigelovii*) in the Eldorado Mountains east of Nelson. This plant likes dry, rocky desert slopes and in the spring displays colorful yellow blossoms at the ends of its upraised arms. Walking through a stand of these can be hazardous as the limb segments littering the ground attach themselves to your shoes, then become scraped off on your shins where they dig into flesh. The painful process goes even further if you grab at the source of pain only to find the cholla segment now embedded in the palm of your hand. *Photo: Joe Tingley*

11d The shoreline of **Lake Mead** northwest of the overlook at **The Cliffs**. Terraces cut in soft sedimentary rocks of the Muddy Creek Formation by wave action mark the fluctuating levels of the lake. *Photo: Joe Tingley*

11e The **beavertail cactus** (*Opuntia basilaris*) is a low (6 to 12 inches) spreading plant that can attain a width of 6 feet. Its jointed stems are oval and flat, resembling a beaver's tail. Bright magenta to red flowers bloom from the top edge of the joints from March to June. *Photo: Steve Castor; inset: Jack Hursh*

PLATE 12

12a One of the water reclamation ponds at the **Henderson Bird Viewing Preserve**. Frenchman Mountain is in the distant background. *Photo: Joe Tingley*

12b A pond at **Tule Springs** with the Las Vegas Range in the background. *Photo: Becky Purkey*

12c **American robin** with lunch. *Photo: Roy W. Cazier*

12d **Fritillary butterflies** are fast flying and are commonly seen as an orange blur as they dart about. *Photo: Roy W. Cazier*

12e Shorebirds on a stopover at the **Henderson Bird Viewing Preserve**. *Photo: Joe Tingley*

12f Radar interferogram map of **subsidence in Las Vegas Valley**, 1992 to 1997, made by comparing satellite radar images taken at different times. Faults are shown in white and the subsidence is shown by the progression of rainbow colors. To find the areas of greatest ground subsidence, start with green-blue at the edge of the map (zero subsidence), move through blue to red to yellow (increasing amount of subsidence), then start over with green-blue, then blue, then red to yellow (most subsidence). The greatest measured subsidence over the 5-year period (about 0.6 feet) is in the center of the bulls eye at the upper left corner of the map. *Credit: Bell and others, 2001*

Trip 5: Water, Subsidence, and Wetlands in Las Vegas Valley

The common thread on this trip is water, and we begin with Corn Creek Springs in the north part of Las Vegas Valley. Visits to Corn Creek and Tule Springs, the latter now being rapidly surrounded by housing developments, provide a look back in time at what Las Vegas Springs, in historical Las Vegas, may have been before urban development and over-pumping of groundwater caused the springs at that site to stop flowing. We then take you to several locations in the urban areas of the valley where you can see faults and the effects of land subsidence due to this groundwater withdrawal.

This journey of water ends with visits to the City of Henderson's Bird Viewing Preserve, developed at the city's water reclamation facility, and to Clark County's Wetlands Park, now under development along Las Vegas Wash. As a sort of postscript, you can take the side trip to Las Vegas Wash, described on Trip 4, to revisit the water flowing from the culvert beneath Lake Las Vegas on its way to Lake Mead—the last local stop for Las Vegas Valley water (it eventually flows on down the Colorado River toward the Gulf of California).

Differing in format from the previous four trips, this trip doesn't follow a set route. Instead, some scenic and geologically interesting spots around Las Vegas Valley are presented and you select from them to plan your own trip.

Most of the locations described are in urban areas of Las Vegas and Henderson and many are at points along streets with very heavy traffic. If you want to more closely examine any of the sites, first find a safe spot to park, then plan your examination. Driving slowly and viewing scenery can be a dangerous combination in Las Vegas traffic.

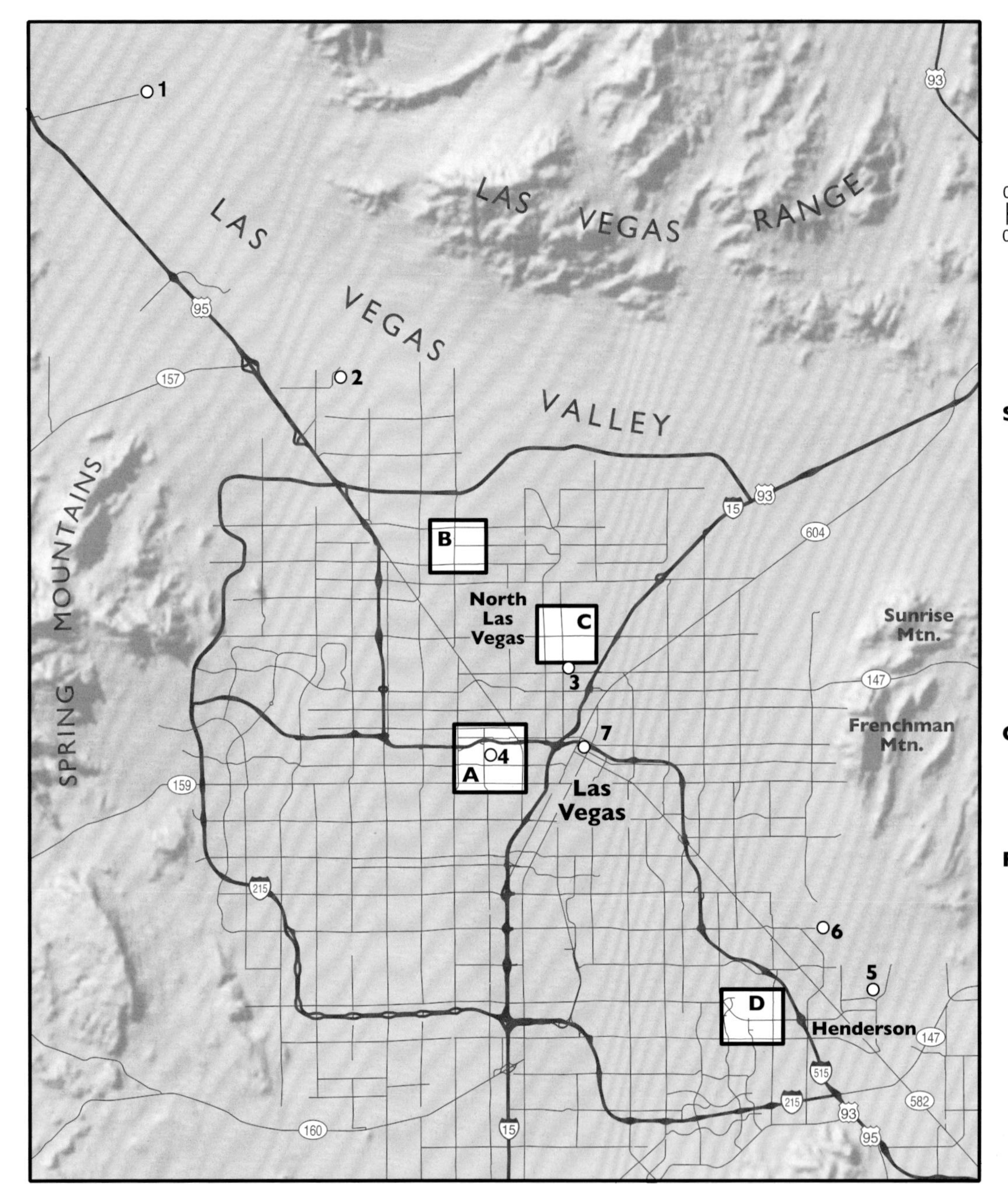

Location map for Trip 5.

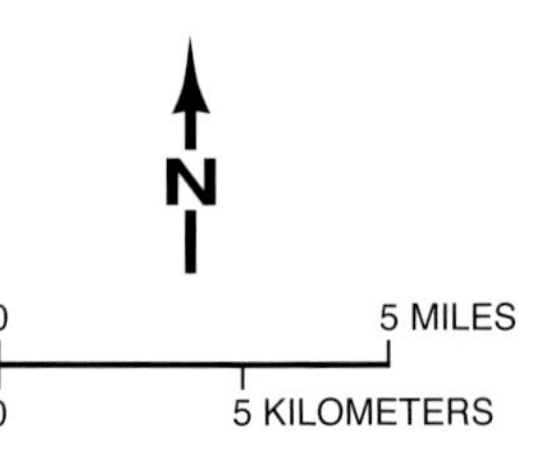

Springs and Wetlands

1. Corn Creek Springs
2. Tule Springs
3. Kyle Ranch Spring
4. Las Vegas Springs
5. Henderson Bird Viewing Preserve
6. Clark County Wetlands Park

Other View Points

7. U.S. Post Office benchmark

Fault Viewing Areas

A. Valley View fault
B. Eglington fault
C. Windsor Park/Cashman Field fault zone
D. Whitney Mesa fault

Natural Springs, The Water Source

Corn Creek Springs

To get to Corn Creek Springs, travel north from Las Vegas on U.S. 95. About 10 miles north of the turnoff to Kyle Canyon (the starting point of Trip 2), look for a sign announcing that the Desert National Wildlife Range is to the right. *(GPS 225)* Take the graded gravel road to the right and follow it for about 4 miles to the Wildlife Range visitor center at Corn Creek Springs. *(GPS 226)*

As you turn onto the gravel road, look for vegetation-covered mounds across the valley beyond the artistically arranged line of mail boxes. These patches of greenery mark the trace of the Corn Creek fault. Now look to the far right and see the low point of Las Vegas Valley defined where the sloping bajada surface extending west from the Sheep Range meets the bajada extending east from the Spring Mountains. Frenchman Mountain, beyond Las Vegas, is visible in the distance at about 1:00.

In the near foreground beyond the mailboxes, notice the thick white Pleistocene spring deposits that cover large portions of Corn Creek Flat. These are the remains of extensive sheets of fine grained sediment deposited from springs along the groundwater-surface interface in the valley bottom when the water table was higher than it is at present. The sediments were deposited during the two most recent glacial periods, one between 25,000 and 14,000 years ago and the other between 14,000 and 8,000 years ago. The wet meadows and marshlands associated with these ancient springs were near the head of Las Vegas Creek, and probably merged with the Tule Springs-marsh-stream area to the southeast that we describe later.

Photo: Becky Purkey

Thick, light-colored Pleistocene spring deposits of Corn Creek Flat.

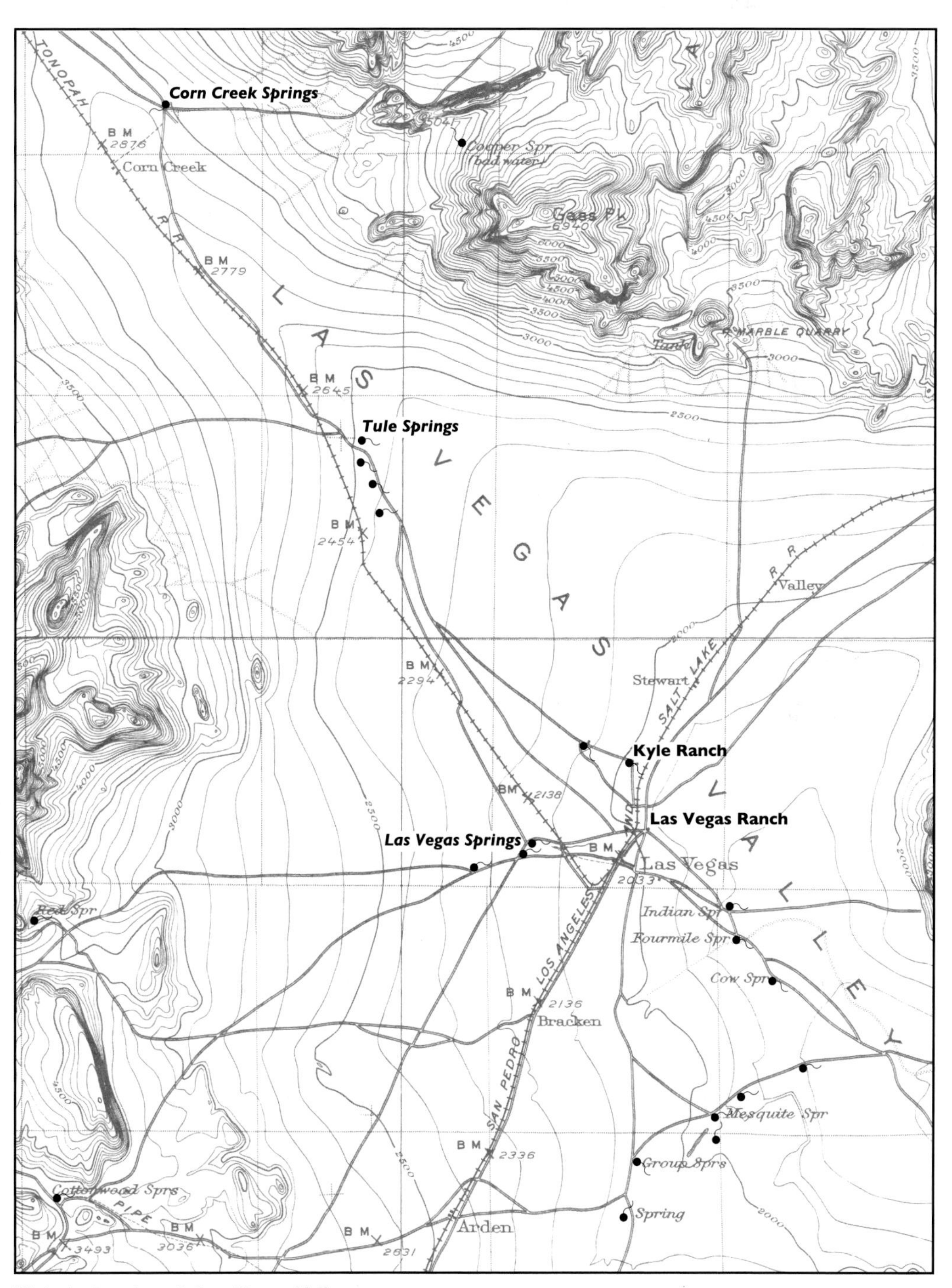

Historical springs in Las Vegas Valley **(shown on U.S. Geological Survey Las Vegas topographic base map, 1908).**

Corn Creek Springs is but one of a series of springs that follow the center of Las Vegas Valley from Indian Springs on the northwest to Las Vegas Springs in what is now the center of Las Vegas on the southeast. The springs are fed by runoff from the high mountains on each side of the valley flowing down into the deep valley gravels where it then moves up along fault structures, perhaps associated with the Las Vegas Valley shear zone, to the surface as springs and seeps.

Water at Corn Creek Springs comes to the surface along the trace of the Corn Creek fault. This fault can be seen just east of the spring ponds near the Wildlife Range visitor center. The vegetation mounds mentioned earlier mark the trace of the fault, and there is a small elevational change across the fault—the bajada surface to the west of the fault has been dropped down.

Stop at the visitor display and sign in, then take time to walk the trails around the spring-fed pond. One trail branch, north of the display area, leads to the right and climbs up across the fault trace. Here you can stand on cemented gravels to the east, and look down upon old ranch buildings situated on down-dropped ground to the west of the fault. Get a good look at the fault scarp in this area, and try to imagine what it would look like if you built a housing project on top of it and, for good measure, paved a road across it. This picture in your mind will be helpful to you when trying to spot similar faults in the heart of Las Vegas at one of the next points on this trip.

There are rest rooms at the display area, and there is a picnic area farther along the gravel road to the east (there are signs). If you have the time, water, and a sturdy vehicle (preferably one with high clearance, good tires, and lots of gas), drive east into the Wildlife Range for exposures of Paleozoic rocks, desert scenery, and maybe even a glimpse of the elusive desert bighorn sheep. Check at the visitor center for information on the roads, however, before you venture into the Range. Also, remember this is a wildlife refuge that is managed much like a wilderness area—there are lots of restrictions including no off-road driving.

Tule Springs

From U.S. 95 North, take Exit 93 at Durango and travel north on Durango for about 1.4 miles, then turn right (east) on Brent Lane another 0.5 miles to reach Tule Springs at Floyd Lamb State Park. There are signs at intervals along the way to lead you to the state park. This area is being rapidly developed, and subdivisions have advanced to the western and southern boundaries of the park. Most of your approach to the park will be through these housing developments. At the park entrance, you will need to pay a day use fee to enter. *(GPS 227)*

Tule Springs today is a large park complex developed around several formerly spring-fed pools (the water table has dropped, the springs no longer flow, and the ponds are maintained by water pumped from wells). There are picnic areas shaded by tall trees, and enough lawn to accommodate almost any kind of recreational activity.

Between about 15,000 and 9,000 years ago, however, this area alternated between stream and marsh environments and, from 32,000 to 11,000 years ago, Tule Springs and its surroundings were inhabited by a variety of animals resembling the present-day wildlife of East Africa. Fossilized remains of large animals including camels, mammoths, jaguars, horses, sloths, and bison have been found in the sediments here. These animals became extinct about 11,000 years ago in southern Nevada—and throughout North America—about the same time that manmade stone projectile points begin to be seen in the archaeological record in North America. At other locations, such as the Lehner site in southern Arizona, stone projectile points have been found along with bones at mammoth kill sites. Here at Tule Springs, scientists also hoped to find evidence of man along with the animals when, in 1962, they excavated and sifted through 200,000 tons of ancient spring deposits. The study, however, was inconclusive.

The mounds of white silty material you see to your left as you drive into the park (also hidden behind the oleanders that rim the parking areas north of the ponds) are calcareous mudstones that were deposited by springs in boggy meadows and marshes that occupied the valley bottom here from about 30,000 to 15,000 years ago during the early and middle Pleistocene glacial epoch. During this time, Las Vegas Valley was drained by a sizeable stream that flowed into Las Vegas Wash. Today, this stream is much smaller and the dynamics of this drainage system are strikingly different due to climate change and the impact of human development in the valley.

Corn Creek fault scarp. Vegetation surrounding the pond at Corn Creek Springs is in the upper left, behind the Desert National Wildlife Range visitor information center. The line of the fault scarp is roughly marked by the bushes to the right of the buildings and by the rise in the road at the right of the photo.

Photos: Joe Tingley

Floyd Lamb State Park at Tule Springs. Gass Peak in the southern Las Vegas Range is in the background.

Springs like Tule Springs provide water for a wide variety of native and introduced plants. Along the hiking and bike trails within the park are signs with the names of various planted trees and shrubs. Among the native desert plants found here are banana yucca, Mojave yucca, Mojave thistle, creosote bush, paperbag bush, indigo bush, buckhorn cholla, strawberry hedgehog cactus, beavertail cactus, spiny menodora, Mormon tea, and catclaw. Other common plants seen include Russian thistle (tumbleweed), oleander, and tamarisk. This latter group is all non-native, introduced here by man. Fish in the spring pools include rainbow trout and channel catfish.

Kyle (Kiel) Ranch Spring and Historical Site

To get to this site, travel north from U.S. 95 on Martin Luther King Jr. Blvd. to Carey Ave. Turn right (east) on Carey and go about one mile to Commerce Street. The ranch site is at the northwest corner of Carey and Commerce, tucked into the corner of a large warehouse complex. Follow Carey beyond the intersection and turn into the entrance to the warehouse, there is a circular paved area where you can park for a short time, get out and read the information on the historical marker. This is the site of one of the historical natural springs around which Las Vegas grew.

Photos: Nevada Historical Society

Mansion built by John S. Park at the old Kiel Ranch. Photo taken in about 1920. The house burned to the ground in 1992.

Kyle (Kiel) Ranch is another example of how geographic names get scrambled with time. Information on the Nevada Historical Marker No. 224 credits Conrad Kiel with establishing the ranch here in 1875. *(GPS 228)* By 1904, the name metamorphosed into "Kyle," and that is the official name for this site (see comments on page 21, Trip 1 about official names). Kyle Canyon in the Spring Mountains (Trip 2) is also named for the Kiel family.

The ranch house burned to the ground in August 1992, and now about all you can do here is take a walk around the perimeter fence and look through at the trees and remaining ranch outbuildings in the grassy low spot.

Las Vegas Springs

The Las Vegas Springs site is located east of South Valley View Blvd. between U.S. 95 on the north and Alta Drive on the south. To get there, take the Valley View Blvd. exit south (Exit 78) from the freeway, drive south on Valley View Blvd. about halfway down the large block to Meadows Lane. At Meadows Lane, turn left into the entrance to the Springs Preserve *(GPS 229)*. Owned by the Las Vegas Valley Water District, this 180-acre parcel of land is being developed into what some believe will be Las Vegas' very own "Central Park." The project, opened in June 2007, includes museums, educational galleries and interactive exhibits, hiking trails, and more than 30,000 plants in an expansive botanical garden. Planned to open in 2009 is a 79,000-square-foot Nevada State Museum.

Las Vegas Spring, about 1900.

Las Vegas Springs, at what is now the main well field of the Las Vegas Valley Water District, is at the historical heart of Las Vegas. These and other nearby springs, such as those at the Kyle (Kiel) Ranch site a little to the north, provided water to "The Meadows"—Las Vegas, in Spanish—that gave their name to the first settlement here, then to the valley and eventually to the city that now dominates the desert landscape.

The first pumping of groundwater logically occurred near the historical natural springs and pumping, since the mid-1930s has progressively lowered the water table. Spring water flowed to the surface as recently as the late 1950s, but increased pumping of groundwater to supply the growing metropolitan area soon put a stop to the natural surface flow and, as pumping here and elsewhere throughout the valley continued, the ground surface around the wellheads and the entire floor of the valley began to compact and settle. At this point, ground subsidence as well as water supply became an urgent issue for the city of Las Vegas. Old well heads on some of the wells now protrude above the ground surface several feet due to subsidence caused by groundwater withdrawal (see photo and sketch on page 126).

Photo: John Bell

The wellhead at Well No. 5 at Las Vegas Valley Water District's main well field (at the site of Las Vegas Springs) is more than 4 feet above the land surface because of subsidence caused by groundwater withdrawal.

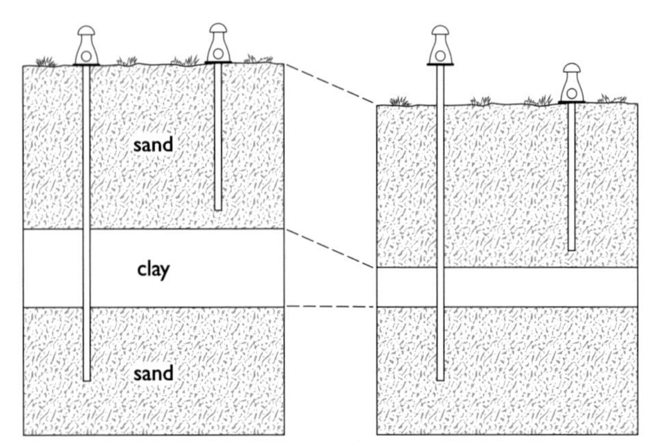

Wellhead protrusion resulting from compaction of soil layers due to groundwater withdrawal (after Mindling, 1965). Wells that do not penetrate the compacted layer are not affected.

SUBSIDENCE—CAUSE AND EFFECT

For decades, municipal and private water wells in Las Vegas Valley have pumped more water from the ground than is naturally returned by precipitation of rain and snow in the higher mountains bordering the valley on the north and west. This precipitation provides most of the groundwater recharge that slowly flows underground through sediments that fill Las Vegas Valley. Most flow occurs through pore space in layers of sandy sediments. These sand or sandstone aquifers are separated by layers of very fine-grained clay. The clay layers slow the penetration of rain water vertically into the ground. In contrast, most of the rain that falls onto the valley floor does not reach the water table. This precipitation quickly runs off into ephemeral streams (as flash flood waters), and much evaporates into the atmosphere.

One adverse result of over-pumping of the groundwater resource is land subsidence. Water can normally be drawn from compacted sand or sandstone aquifers without subsidence, but when clays occur in the sediments, as they do in sediments in the central and eastern part of the valley, removal of water allows the clays to compress and the overlying land surface to sink.

Downtown Las Vegas has actually dropped in elevation by more than 5 feet since 1935, when geodetic monitoring in the valley was begun. Much of the valley is sinking at a relatively uniform rate so that most structures are not adversely affected. Locally, however, the subsidence is focused on preexisting geologic faults, which serve as points of weakness for ground movement. Geodetic control lines placed for ten years (1978–1987) over several selected faults in areas of high subsidence revealed movement of as much as 2 inches per year on several of the faults.

Land subsidence in Las Vegas Valley can be attributed to tectonic activity, natural prehistoric dewatering of the basin sediments, increased loading from the weight of Lake Mead (measurements made from 1935 to 1950 show that the impounded waters have caused a depression of about 7 inches centered on the lake and 4 to 5 inches of tilting of the Las Vegas Valley), and historic groundwater extraction. Of these potential causes, the last—man's influence on the groundwater system—is, without a doubt, the most important and immediate factor (and the only controllable factor).

In the absence of geodetically controlled leveling data, subsidence-induced movement on the faults can be detected by the development of cracks or fissures in the ground that form on or in close proximity to the faults. The fissures begin as hairline cracks which then are widened by surface water that erodes the walls of the crack. These thin cracks are deep. They originate at depths of perhaps tens of feet and migrate upward with time. Finally they cut the land surface. Some fissures are hundreds to thousands of feet long and are widened by erosion at the surface to as much as several feet.

Houses built over these cracks and subsequent erosion gullies can be severely damaged; foundations crack and tilt. As much as $12 million in damage has occurred in

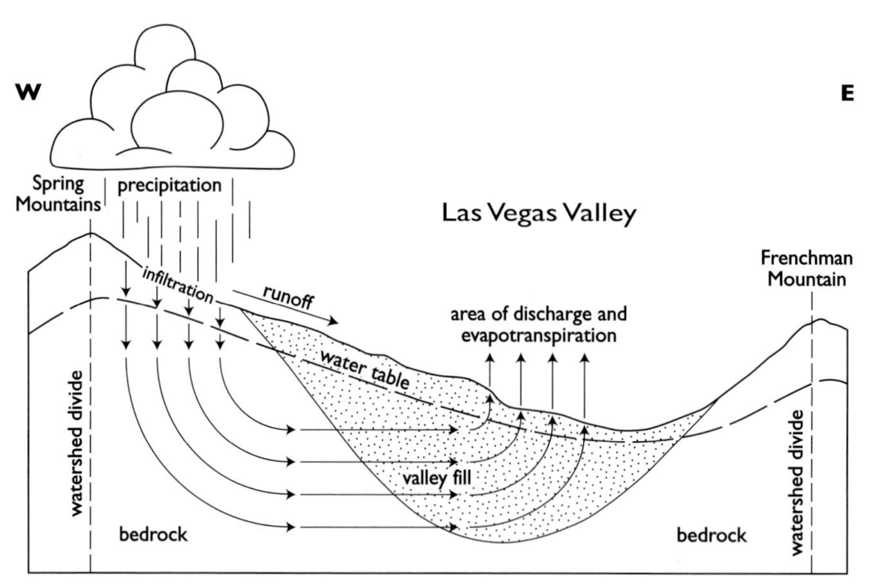

Generalized groundwater flow system in Las Vegas Valley (after Domenico and others, 1964).

Photo: Jon Price

Fissure, widened by erosion passing beneath the foundation of an abandoned home near Simmons Street in North Las Vegas.

one subdivision in North Las Vegas. Most of the fissures that have developed in Las Vegas Valley occur near faults (within 1,200 feet). Ways to avoid this geological hazard include not building over known fissures or near faults, limiting irrigation in the areas of known fissures, and channeling rain and irrigation waters away from fissures.

Beginning in the early 1990s, the Las Vegas Valley Water District, the major water supplier in the valley, has been artificially recharging the aquifers at depth. Colorado River water (from Lake Mead) is injected into the Las Vegas aquifer during the winter months, then pumped back out during the high-demand summer months for irrigation of lawns and golf courses. This recharge program has arrested subsidence throughout much, but not all of the valley.

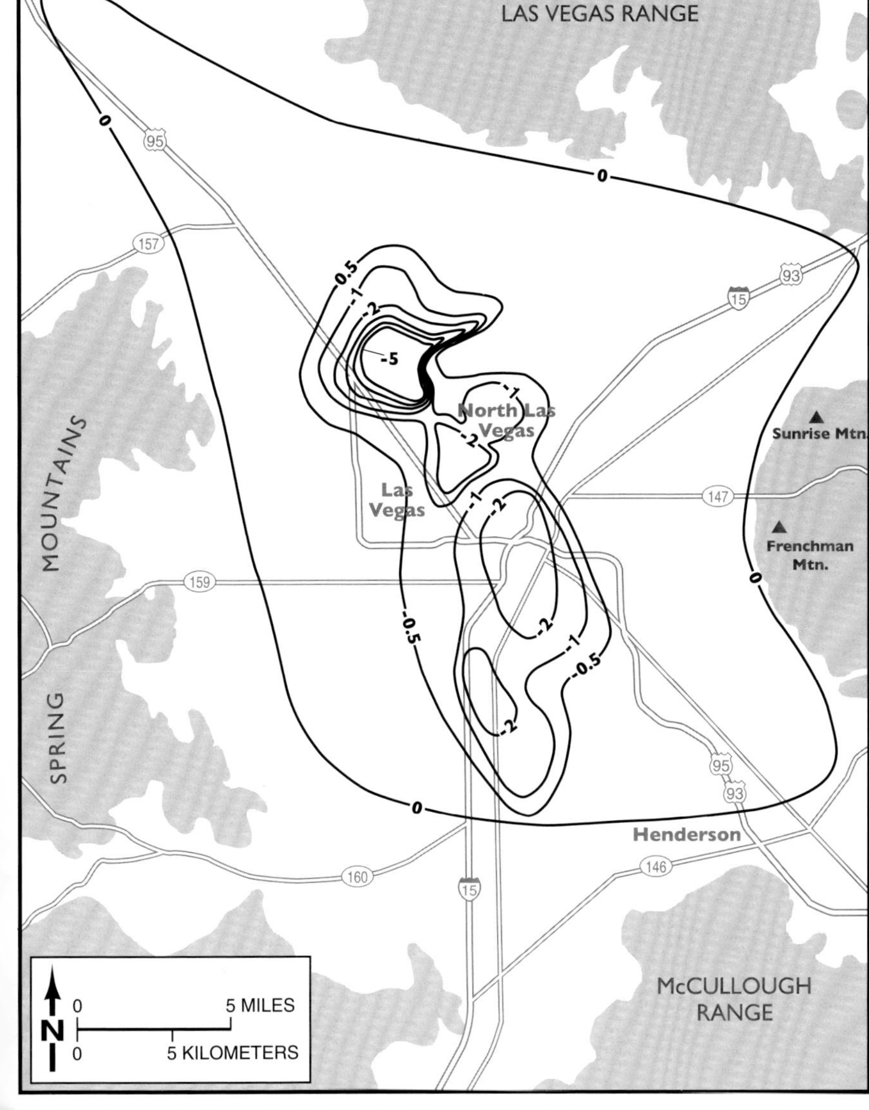

Land surface subsidence (in feet) in Las Vegas Valley between 1963 and 2000 (after Bell and others, 2001).

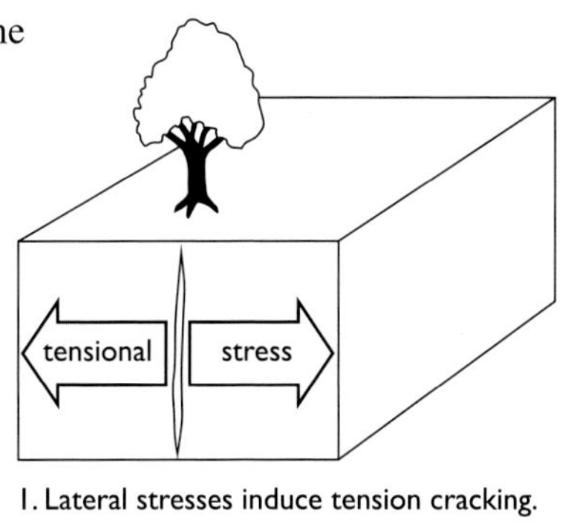

1. Lateral stresses induce tension cracking.

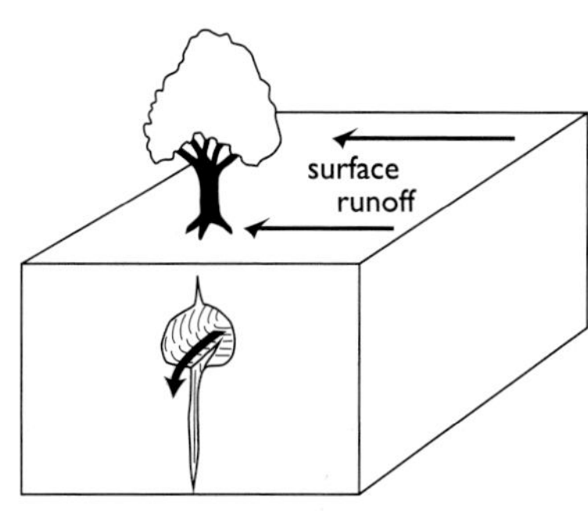

2. Surface runoff and infiltration enlarge crack through subsurface piping.

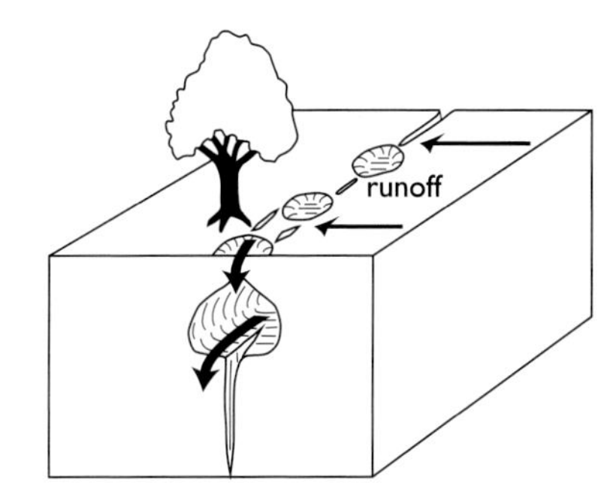

3. As piping continues, the fissure begins to appear at surface as series of potholes and small cracks.

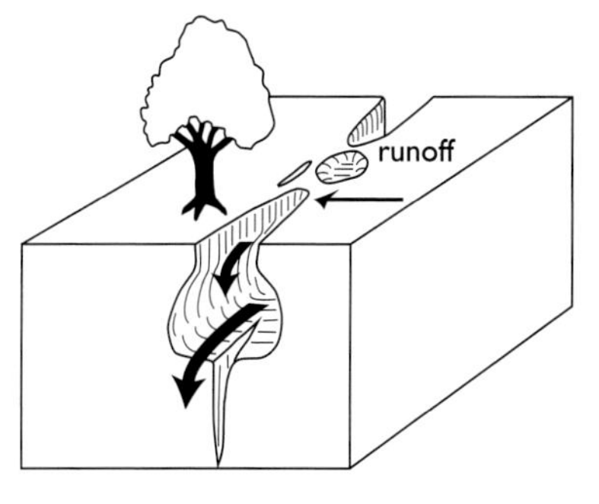

4. As infiltration and erosion continue, the fissure enlarges and completely opens to the surface as the tunnel roof collapses.

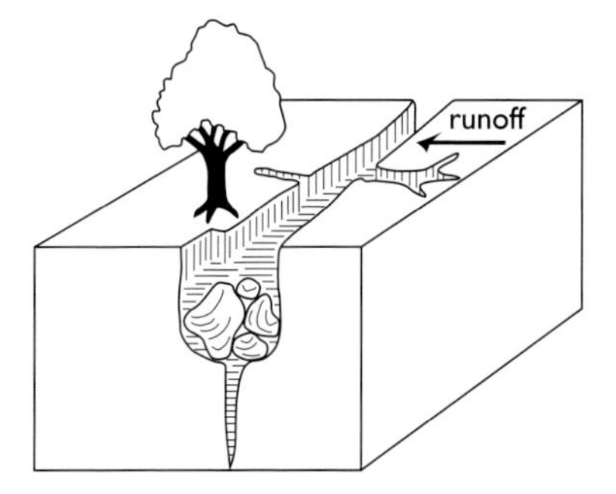

5. The entire fissure is opened to the surface and enlargement continues as the fissure walls are widened; extensive slumping and side-stream gullying occur.

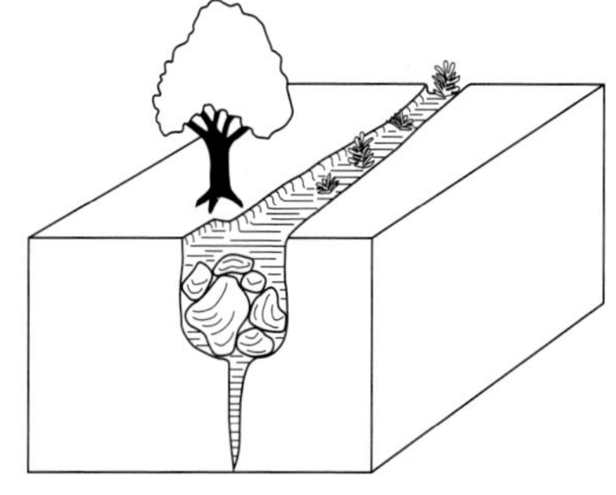

6. The fissure becomes filled with slump and runoff debris and is marked by a line of vegetation and slight surface depression.

Development of fissures (after Bell, 1981).

U.S. Post Office Benchmark

This survey benchmark is on the right (southeast) side of the steps *(GPS 230)* of the old U.S. Post Office building at the intersection of Stewart Ave. and Third St. —within walking distance of hotels and casinos in downtown Las Vegas. No longer a Post Office, the building has been acquired by the city for future use as a downtown cultural facility. The benchmark was surveyed in 1936. Since that time, the benchmark, the Post Office, and the surrounding land have subsided about 5 feet relative to presumably stable benchmarks in bedrock in the mountain ranges.

Until recently, surveying benchmarks was tedious work that required the use of transits or other telescope-type instruments and tape measures. New technology, including Global Positioning System (GPS) satellites and radar interferometry, has changed the way subsidence is measured.

GPS surveying is done using a receiver on the ground that picks up a signal from each of several Earth-orbiting satellites overhead. Precise clocks in the satellites allow measurement of the distance from the satellite to the ground (again, by measuring time and calculating distance from the speed of light or radio signals). By using two receivers, precise locations can be calculated to within fractions of an inch.

The most recent technology, however, is radar interferometry, a powerful new tool that uses radar signals from satellites to measure changes in the land surface. By comparing radar scans made of the same area at different times, it is possible to make high-precision measurements of land surface elevation. Under ideal conditions, elevation changes of 0.4 inches or less can be detected.

Finding Faults

For this part of the trip, we take you to several locations up and down the valley where you can see fault scarps crossing or perhaps running parallel to the road. These scarps mark the surface locations of faults that cut the alluvial sediments that fill Las Vegas Valley.

When rock masses are stressed due to forces operating in the Earth's crust and the stress exceeds the rock's ability to stay in one piece, the rock fails or breaks and moves in whatever direction it can to relieve the stress. The movement of rock along this break causes an earthquake, and the plane along which the rocks move is called a fault. Erosion gradually smoothes the faulted or broken surface to form a gentle scarp that slopes down in the direction of movement (remember the fault scarp at Corn Creek Springs).

These faults are interesting because they are sites of past and possibly future earthquakes. There are a number of active or potentially active faults in and near Las Vegas Valley. Earthquakes of magnitude 3 or smaller are fairly common and generally cause no damage. Significant earthquakes, of magnitude 6 or greater, are much less frequent, and may occur on a given fault only once every few thousand years.

Subsidence cracks and faults have been observed and mapped throughout Las Vegas Valley. Many of these features, easy to observe only a few years ago, have now been filled, covered by subdivisions, or paved over and are now difficult or impossible to see. There are, however, several areas along the valley where major streets cross these structures and there is a bump, sag, or a visible elevation change that can be seen or even felt as you drive across it. Drive to one or two of the locations on the following list and see how good you are at finding faults.

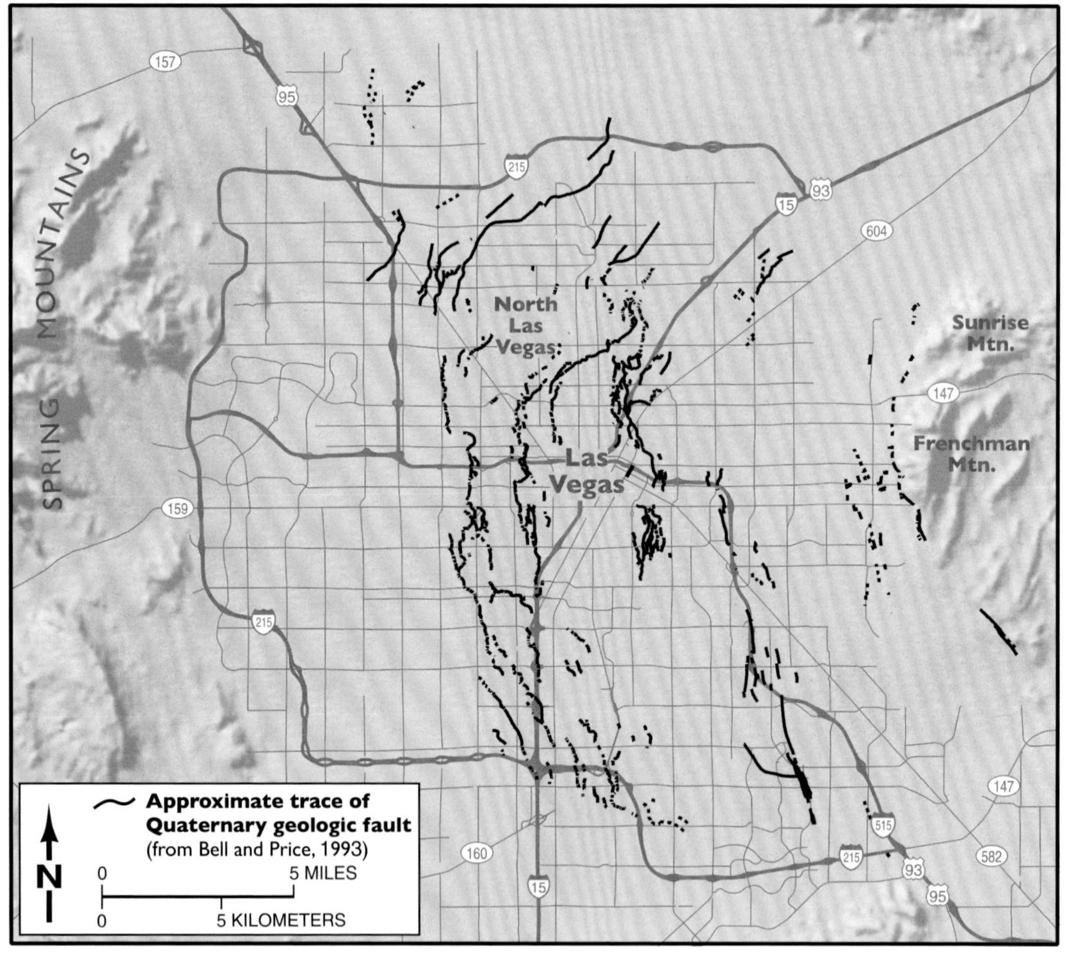

Major faults in Las Vegas Valley.

Valley View Blvd. at Charleston Blvd.

This location is in front of the Las Vegas Valley Water District office. Driving north on Valley View Blvd. north from Charleston, notice that the Water District office building is situated on ground below you to the east, while the area to the west of Valley View is high. The Valley View fault parallels Valley View Blvd. here, and the ground to the east has subsided. *(GPS 231)* This fault extends north of Alta, swings to the east into the well field at Las Vegas Springs, and extends north across U.S. 95 where it runs north parallel to the west side of Lorenzi Park. When you enter the park from the west, you drive down across the fault scarp to visit the Nevada Historical Society, picnic area, and playing fields located on the downdropped ground to the east.

The approximate trace of the Valley View fault between Charleston Blvd. and Lorenzi Park. ▼

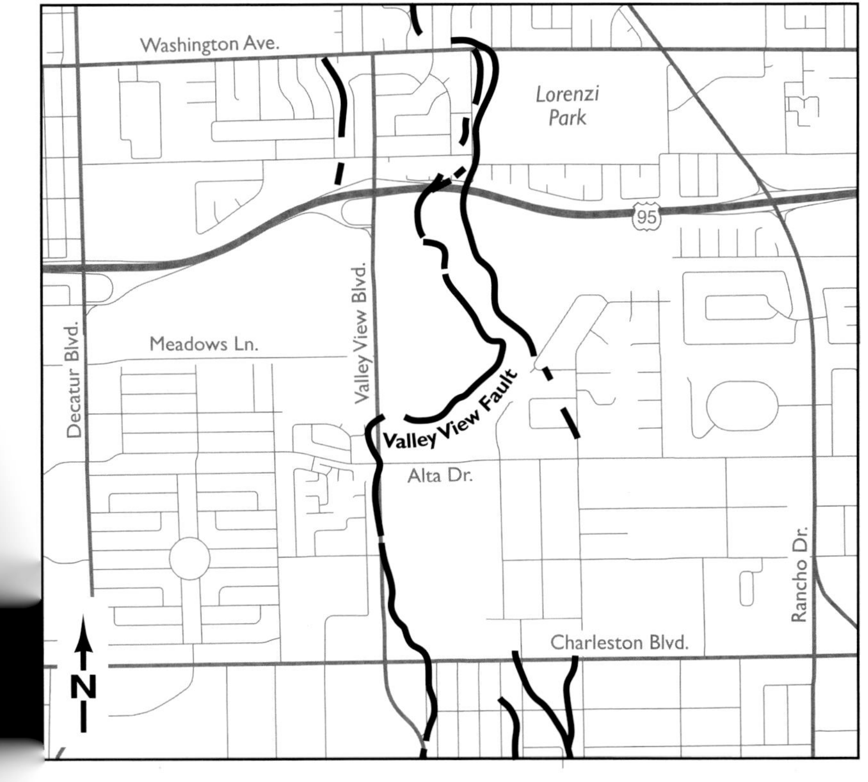

Ann Road at Decatur Blvd.

Ann Road crosses the scarp of the Eglington fault just east of its intersection with Decatur. As you drive east on Ann Road, you will feel the pavement drop to the east in two separate, gentle rolls.

Cheyenne Avenue and Revere Street

Heading east on Cheyenne just east of Revere, the road drops across the scarp of the Windsor Park/ Cashman Field fault zone. *(GPS 232)* The large school bus parking area to the east is on the subsided ground. The fault trace heads northeast from here toward the recycling plant and southwest around the white bluff west of the buses.

The approximate trace of the Eglington fault near Decatur Blvd. and Ann Road. ▶

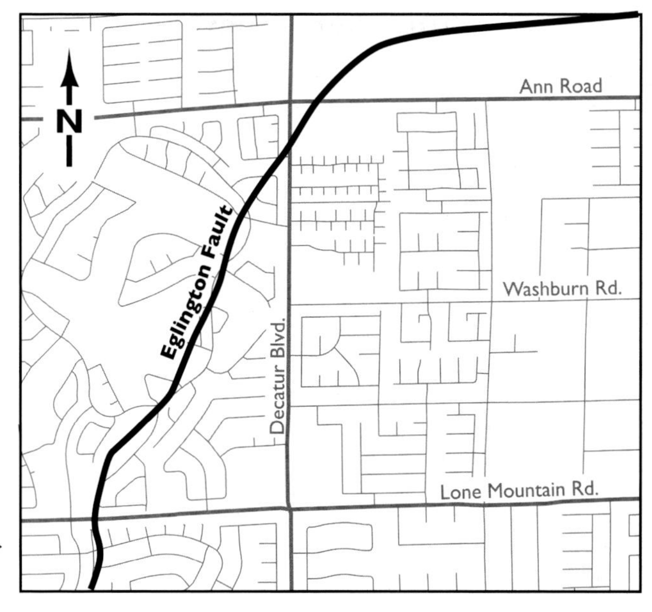

The approximate trace of the Windsor Park/ Cashman Field fault zone near Cheyenne Ave. and Revere Street. ▶

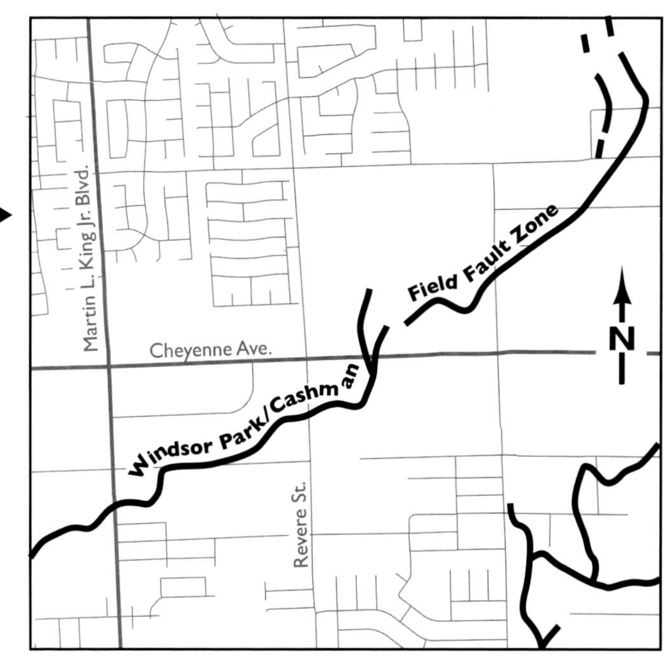

Sunset Road west of Arroyo Grande Blvd.

This area is south of Whitney Mesa, in the south end of the valley. To get there, take the Sunset Road West exit from I-515. Follow Sunset Road west as it curves up along the south side of Whitney Mesa. Just west of the intersection with Arroyo Grande Blvd., as the road curves to the right, notice a slight dip in the pavement as you cross the Whitney Mesa fault. The cement curb here also shows a sag across the fault zone. *(GPS 233)*

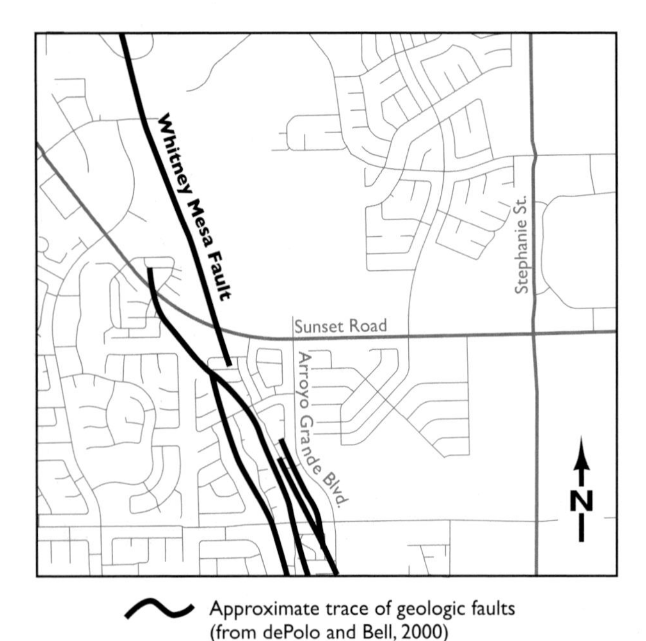

Approximate trace of the Whitney Mesa fault near Sunset Road.

Downstream

The City of Henderson Water Reclamation Facility and Bird Viewing Preserve

This is almost the end of the trail for Las Vegas Valley water used and discarded by the City of Henderson. Wastewater treatment plants are usually hidden away at the edge of cities and are avoided by sensitive residents. Henderson, however, has taken a positive step and transformed more than 90 acres of its water reclamation lagoons, ponds, and infiltration basins into a protected sanctuary for birds and wildlife. Both birds and birders have been attracted to the water treatment area since the 1960s (it is claimed to be the third largest body of water in southern Nevada) and the ponds proved irresistible to a wide variety of native and migratory birds. According to the fact sheet available at the visitors center, the Henderson Bird Viewing Preserve is the "Place to Call Home" for almost 200 species of birds, including waterfowl, wading birds, and birds of prey. The birding program at the Water Reclamation Facility officially began in 1995, making it one of only a few "dual use" facilities in the United States. Vegetation is allowed to flourish around the margins of the ponds providing both resting and nesting areas for the birds. For the bird and nature lovers, there are landscaped paths with observation stations, informational signs, and scattered benches.

To get there, take the Sunset Road exit east from I-515, or turn east on Sunset Road from the Boulder Highway. Turn north on Moser Drive and follow it to Athens Ave., jog a little to the east and follow the signs into the Bird Viewing Preserve parking area. You will need to go inside and register, but there is no fee and the staff is helpful and more than willing to give you pointers on birds. The preserve is open daily from 6:00 a.m. to 3:00 p.m. *(GPS 234)*

Photo: Joe Tingley

One of the bird viewing ponds at the Henderson Water Reclamation Facility. Frenchman Mountain is in the background.

Clark County Wetlands Park

Water from the entire 1,600-square-mile watershed of Las Vegas Valley passes along the lower reaches of Las Vegas Wash and through the Clark County Wetlands Park on its journey out of the valley and on to Lake Mead. Presently, the average flow through the wash is 153 million gallons of water per day. This flow includes highly treated waste water from the Clark County Sanitation District, the City of Las Vegas, and the City of Henderson, along with shallow groundwater, urban runoff, and stormwater runoff.

For 1,000 to 4,000 years prior to white settlement of Las Vegas Valley, the wash was apparently dry. As the valley began to urbanize, treated wastewater and urban runoff entered the wash, increased the flow in its channel, and a wetlands area began to develop on the low eastern side of the valley. In addition to functioning as a settlement and natural purification area for the water, the wetlands provided habitat for numerous desert plants and animals and, by the 1970s, Las Vegas Wash supported more than 2,000 acres of wetlands. However, as the urban population increased so did the flow of water in Las Vegas Wash increase. Many factors contributed to this, including increased output of water from treatment facilities and increased surface runoff during flash floods (as more areas in the valley are paved with asphalt and concrete, less water percolates into the ground and more runs off into Las Vegas Wash). This increased runoff caused erosion of the channel of Las Vegas Wash, and destroyed much of the wetlands created earlier. Today there are less than 400 acres of wetlands along Las Vegas Wash.

Clark County Wetlands Park was conceived as a wash stabilization and erosion control project to stop and maybe help reverse the damage being done by the ever-increasing water flow into Las Vegas Wash. The welcome fallout from this venture is restoration of wildlife habitat and wetland vegetation along the wash. The master plan, put into place following passage of a $13.3 million wildlife and parks bond in 1991, calls for the construction of erosion control structures, horse, bike, and walking trails, a nature preserve, a visitor center, and a small amphitheater. The park will not be completed until 2015 but, for now, some of the trails are open, and there are areas suitable for picnicking.

The Wetlands Park is reached by driving east from the Boulder Highway on either Tropicana Ave. (to approach on the north) or on Russell Road (to approach on the south). Trailheads and the visitor center are in the park just beyond the end of Tropicana Ave. If you approach from Russell Road, turn north on Broadbent Blvd. (just before you reach Sam Boyd Stadium) and follow Broadbent Blvd. to Tropicana Ave., then turn right (east) to the park. The Wetlands Park is open daily from 9 am to 5 pm. *(GPS 235)*

POSTSCRIPT

See the "Side Trip to Las Vegas Wash" described on page 94, Trip 4, for a description of the lower part of Las Vegas Wash, below the Lake Las Vegas development.

Photo: Marguerite Gregory, California Academy of Sciences

American avocets (*Recurvirostra americana*) are common summer residents in Nevada. These long-legged shorebirds are black and white above and white below. The head and neck are a rusty cinnamon color during breeding season, whitish gray in winter. Avocets can be seen in marshes, shallow lakes, and alkali ponds sweeping their long, up-curved bills from side to side in search of small crustaceans and aquatic insects. ▶

Photo: Dr. Lloyd Glenn Ingles, California Academy of Sciences

◀ **The western grebe (*Aechmophorus occidentalis*) is only found in western North America. It is black and white, with a long neck, red eyes, and a long, pointed bill. These birds are superbly adapted swimmers and divers whose principal diet consists of small fish and aquatic insects. Chicks don't dive well and are often carried on their parents' backs for their first several weeks of life.**

BIBLIOGRAPHY

Adams, G.F., and Wyckoff, J., 1971, Landforms: Golden Press, New York, 160 p.

Allen, G.L., Jacobs. J.H., and Hunter, J.W., 1945, Utilization of Three Kids manganese ore in the production of electrolytic manganese: U.S. Bureau of Mines Report of Investigations 3815, 579 p.

Amelung, F., Galloway, D.L., Bell, J.W., Zebker, H.A., and Laczniak, R.J., 1999, Sensing the ups and downs of Las Vegas: InSAR reveals structural control of land subsidence and aquifer-system deformation: Geology, v. 27, no. 6, p. 483–486.

Anderson, R.E., 1973, Large-magnitude late Tertiary strike-slip faulting north of Lake Mead, Nevada: U.S. Geological Survey Professional Paper 794, 18 p.

Anderson, R.E., 1977, Geologic map of the Boulder City 15-minute quadrangle, Clark County, Nevada: U.S. Geological Survey Map GQ-1395.

Anderson, R.E., 2003, Geologic map of the Callville Bay Quadrangle, Clark County, Nevada and Mohave County, Arizona: Nevada Bureau of Mines and Geology Map 139.

Armentrout, J.M., Cole, M.R., and Terbest, H., Jr., eds., 1979, Cenozoic paleogeography of the western United States—Pacific Coast Paleogeography Symposium 3: Society of Economic Paleontologists and Mineralogists, Pacific Section, 335 p.

Armstrong, R.L., 1970, Geochronology of Tertiary igneous rocks, eastern Basin and Range province, western Utah, eastern Nevada and vicinity, U.S.A.: Geochimica et Cosmochimica Acta, v. 34, p. 203–232.

Atkin, B.C., and Johnson, J.A., 1988, The Earth—Problems and perspectives: Blackwell Scientific Publications, Palo Alto, Calif., 452 p.

Axen, G.J., 1984, Thrusts in the eastern Spring Mountains, Nevada—Geometry and mechanical implications: Geological Society of America Bulletin, v. 95, p. 1202–1207.

Axen, G.J., 1986, Thrust faults and synorogenic conglomerates in the southern Spring Mountains, *in* Field trip guide to the geology of southern Nevada: National Association of Geology Teachers, Far West Section Meeting, Oct. 3–5, 1986, Las Vegas, Nev., p. 16–22.

Axen, G.J., 1987, The Keystone thrust and Red Spring thrust faults in the La Madre Mountain area, eastern Spring Mountains, Nevada: Geological Society of America Centennial Field Guide, v. 1, Cordilleran Section, p. 57–60.

Axen, G.J., 1989, Reinterpretations of the relations between the Keystone, Red Springs, Contact, and Cottonwood faults; eastern Spring Mountains, Clark County, Nevada—Discussion: The Mountain Geologist, v. 26, no. 3, p. 69–70.

Bates, R.L., and Jackson, J.A., eds., 1995, Glossary of geology CD-ROM, third edition: American Geological Institute, Alexandra, Va.

Bell, J.W., 1978, Geologic constraints map of the Las Vegas SE Quadrangle: Nevada Bureau of Mines and Geology Map 3Am.

Bell, J.W., 1981, Subsidence in Las Vegas Valley: Nevada Bureau of Mines and Geology Bulletin 95, 84 p.

Bell, J.W., and Price, J.G., 1993, Subsidence in Las Vegas Valley, 1980–91, final project report: Nevada Bureau of Mines and Geology Open-File Report 93-4, 182 p.

Bell, J.W., and Smith, E.I., 1980, Geological map of the Henderson Quadrangle, Clark County, Nevada: Nevada Bureau of Mines and Geology Map 67.

Bell, J.W., Blewitt, G., and Amelung, F., 2001, Las Vegas Valley Subsidence Report, Nevada Bureau of Mines and Geology Open-File Report 01-5, 45 p.

Bell, J.W., Ramelli, A.R., and Caskey, S.J., 1998, Geologic map of the Tule Springs Park Quadrangle, Nevada: Nevada Bureau of Mines and Geology Map 113.

Bell, J.W., Ramelli, A.R., and dePolo, C.M., 2001, 1998 subsidence report, prepared for Las Vegas Valley Water District: Nevada Bureau of Mines and Geology Open-File Report 01-04, 123 p.

Bell, J.W., Ramelli, A.R., dePolo, C.M., Maldonado, F., and Schmidt, D.L., 1999, Geologic map of the Corn Creek Springs Quadrangle, Nevada: Nevada Bureau of Mines and Geology Map 121.

Bezy, J.V., 1978, A guide to the desert geology of Lake Mead N.R.A.: Southwest Parks and Monuments Association, Globe, Ariz., 67 p.

Bingler, E.C., 1977, Geologic map of the Las Vegas SE Quadrangle: Nevada Bureau of Mines and Geology Map 3Ag.

Bingler, E.C., and Bonham, H.E., 1973, Reconnaissance geologic map of the McCullough Range and adjacent areas, Clark County, Nevada: Nevada Bureau of Mines and Geology Map 45.

Bissell, H.J., 1969, Permian and Lower Triassic transition from shelf to basin (Grand Canyon, Arizona to Spring Mountains, Nevada), *in* Baars, D.L., ed., Geology and natural history of the Grand Canyon region: Four Corners Geological Society Guidebook, Fifth Field Conference, p. 135–169.

Bohannon, R.G., 1983, Mesozoic and Cenozoic tectonic development of the Muddy, North Muddy, and northern Black Mountains, Clark County, Nevada, *in* Miller, D.M., Todd, V.R., and Howard, K.A., eds., Tectonic and stratigraphic studies in the eastern Great Basin: Geological Society of America Memoir 157, p. 125–148.

Bohannon, R.G., 1984, Nonmarine sedimentary rocks of Tertiary age in the Lake Mead region, southeastern Nevada and northwestern Arizona: U.S. Geological Survey Professional Paper 1259, 72 p.

Bohannon, R.G., and Bachhuber, F., 1979, Road log from Las Vegas to Keystone thrust area and Valley of Fire via Frenchman Mountain, *in* Newman, G.W., and Goode, H.D., eds., Basin and Range Symposium and Great Basin Field Conference: Rocky Mountain Association of Geologists and Utah Geological Survey, 1979, p. 579–596.

Bohannon, R.G., and Morris, R.W., 1983, Geology and mineral resources of the Red Rocks Escarpment Instant Study Area, Clark County, Nevada: U.S. Geological Survey Map MF-1522.

Brenner, E.F., and Glanzman, R.K., 1979, Tertiary sediments in the Lake Mead area, Nevada, *in* Newman G.W., and Goode, H.D., eds., Basin and Range Symposium and Great Basin Field Conference: Rocky Mountain Association of Geologists and Utah Geological Survey, 1979, p. 313–323.

Burchfiel, B.C., and Davis, G.A., 1988, Mesozoic thrust faults and Cenozoic low-angle normal faults, eastern Spring Mountains, Nevada and Clark Mountains thrust complex, California, *in* Weide, D.L., and Faber, M.L., eds., This extended land—Geological journeys in the southern Basin and Range: Field Trip Guidebook, Geological Society of America, Cordilleran Section Meeting, Las Vegas, Nev., 1988, p. 87–106.

Burchfiel, B.C., and Royden, L.H., 1984, The Keystone thrust fault at Wilson Cliffs, Nevada is not the Keystone thrust—Implications: Geological Society of America Abstracts with Programs, v. 16, no. 6, p. 458.

Burchfiel, B.C., Fleck, R.J., Secor, D.T., Vincelette, R.R., and Davis, G.A., 1974, Geology of the Spring Mountains, Nevada: Geological Society of America Bulletin, v. 85, p. 1013–1022.

Campagna, D.J., and Aydin, A., 1991, Tertiary uplift and shortening in the Basin and Range; the Echo Hills, southeastern Nevada: Geology, v. 19, p. 485–488.

Carlson, H.S., 1974, Nevada place names-A geographical dictionary: University of Nevada Press, Reno, 282 p.

Carr, M.D., Evans, K.V., Fleck, R.J., Frizzell, V.A., Ort, K.M., and Zartman, R.E., 1986, Early Middle Jurassic upper limit for movement on the Keystone thrust, southern Nevada: Geological Society of America Abstracts with Programs, v. 18, no. 5, p. 345.

Carr, M.D., McDonnell-Canan, C., and Weide, D.L., 2000, Geologic map of the Blue Diamond SE Quadrangle, Nevada: Nevada Bureau of Mines and Geology Map 123.

Castor, S.B., Faulds, J.E., Rowland, S.M., and dePolo, C.M., 2000, Geologic map of the Frenchman Mountain Quadrangle, Clark County, Nevada: Nevada Bureau of Mines and Geology Map 127.

Clark, J.L., 1993, Nevada wildlife viewing guide: Falcon Press Publishing Co., Inc., Helena and Billings, Montana, 87 p.

Cleveland, G.B, 1975, The flash flood at Nelsons Landing, Clark County, Nevada: California Geology, v. 28, p. 51–56.

Cline, G.G., 1963, Exploring the Great Basin: University of Nevada Press, Reno, 254 p.

Davis, G.A., 1973, Relations between the Keystone and Red Spring thrust faults, eastern Spring Mountains, Nevada: Geological Society of America Bulletin, v. 84, p. 3709–3716.

dePolo, C.M., and Bell, J.W., 2000, Map of faults and Earth fissures in the Las Vegas area: unpublished map, Nevada Bureau of Mines and Geology files.

Dohrenwend, J., 1989, Rates and patterns of piedmont evolution in the southwest Basin and Range, *in* Late Cenozoic evolution of the southern Great Basin: Nevada Bureau of Mines and Geology Open-File Report 89-1, p. 151–152.

Domenico, P.A., Stephenson, D.A., and Maxey, G.B., 1964, Groundwater in Las Vegas Valley: University of Nevada, Desert Research Institute Technical Report, no. 7, 53 p.

Drobeck, P.A., Kern, R.R., Jenkins, S.L., McClure, D.L., Ausburn, K., Linder, E.J., Huskinson, E.J., Hillemeyer, F.L., and Dodge, W.T., 1988, Gold deposits of the Las Vegas region, *in* Weide, D.L., and Faber, M.L., eds., This extended land-Geological journeys in the southern Basin and Range: Field Trip Guidebook, Geological Society of America, Cordilleran Section Meeting, Las Vegas, Nev., 1988, p. 65–86.

Duebendorfer, E.M., 2003, Geologic map of the Government Wash Quadrangle, Nevada: Nevada Bureau of Mines and Geology Map 140.

Duebendorfer, E.M., and Black, R.A., 1992, The kinematic role of transverse structures in continental extension: An example from the Las Vegas Valley shear zone, Nevada: Geology, v. 20, p. 1107–1110.

Duebendorfer, E.M., and Simpson, D.A., 1994, Kinematics and timing of Tertiary extension in the western Lake Mead region, Nevada: Geological Society of America Bulletin, v. 106, no. 8, p 1057–1073.

Duebendorfer, E.M., and Wallin, E.T., 1991, Basin development and syntectonic sedimentation associated with kinematically coupled strike-slip and detachment faulting, southern Nevada: Geology, v. 19, p. 87–90.

Duebendorfer, E.M., Beard, L.S., and Smith, E.I., 1998, Restoration of Tertiary deformation in the Lake Mead region, southern Nevada: The role of strike-slip transfer zones, *in* Faulds, J.E., and Stewart, J.H., eds., Accommodation zones and transfer zones: The regional segmentation of the Basin and Range Province: Geological Society of America Special Paper 323, p 127–148.

Duebendorfer, E.M., Sewall, A.J., and Smith, E.I., 1990, The Saddle Island detachment: An evolving shear zone in the Lake Mead area, Nevada, *in* Wernicke, B.P., ed., Basin and Range extension near the latitude of Las Vegas, Nevada: Geological Society of America Memoir 176, p. 77–97.

Duebendorfer, E.M., Smith, E.I., and Faulds, J.E., 1993, Introduction to the area north of I-40: between Lake Mead, Nevada and Needles, California, *in* Sherrod, D.R., and Nielson, J.E., eds., Tertiary stratigraphy of highly extended terranes, California, Arizona, and Nevada: U.S. Geological Survey Bulletin 2053.

Earl, P.I., 1986, This was Nevada: Nevada Historical Society, Reno, 192 p.

Evans, D.B., 1971, Auto tourguide to the Lake Mead National Recreation Area: Southwest Parks and Monuments Association, Globe, Ariz., 39 p.

Feuerbach, D.L., and Smith, E.I., 1987, Late-Miocene Fortification Hill basalt, Lake Mead area, Nevada and Arizona—Source areas and conduit geometry: Geological Society of America Abstracts with Programs, v. 19, no. 6, p. 376–377.

Fiero, B., 1986, Geology of the Great Basin: University of Nevada Press, Reno, 198 p.

Fiero, G.W., 1976, Nevada's Valley of Fire: KC Publications, Las Vegas, 32 p.

Friess, S., 2007, Las Vegas Springs Preserve brings the wild to Sin City, latimes.com, June 15, 2007.

Gilluly, J., Waters, A.C., and Woodford, A.O., 1959, Principles of geology: W.H. Freeman and Co., San Francisco, 435 p.

Gans, P.B., Landau, B., and Darvall, P., 1994, Ashes, ashes, all fall down: Caldera-forming eruptions and extensional collapse of the Eldorado Mountains, southern Nevada: Geological Society of America Abstracts with Programs, v. 26, no. 2, p. 53.

Glancy, P.A., and Whitney, J.W., 1986, Las Vegas Wash—Dynamic evolution of a southern Nevada drainage channel: Geological Society of America Abstracts with Programs, v. 18, no. 6, p. 615.

Glass, M., and Glass, A., 1981, Touring Nevada: University of Nevada Press, Reno, 253 p.

Hamblin, W.K., 1975, The Earth's dynamic systems—A textbook in physical geology: Burgess Publishing Co., Minneapolis, Minn., 578 p.

Hansen, M.W., 1979, Crinoid shoals and associated environments, Mississippian of southern Nevada, *in* Newman, G.W., and Goode, H.D., eds., Basin and Range Symposium and Great Basin Field Conference: Rocky Mountain Association of Geologists and Utah Geological Survey, 1979, p. 259–266.

Harrill, J.R., 1976, Pumping and ground-water storage depletion in Las Vegas Valley, Nevada, 1955–74: Nevada Department of Conservation and Natural Resources, Water Resources Bulletin 44, 70 p.

Harrill, J.R., and Katzer, T., 1980, Groundwater map of the Las Vegas SE Quadrangle: Nevada Bureau of Mines and Geology Map 3Af.

Harrington, M.R., 1933, Gypsum Cave, Nevada: Southwest Museum Papers, no. 8, 197 p.

Harrington, M.R., 1937, Ancient tribes of the Boulder Dam country: Southwest Museum Leaflets, no. 9, 13 p.

Harrison, C., and Greensmith, A., 1993, Birds of the world: Dorling Kindersly, Inc., New York, 416 p.

Haynes, C.V., 1967, Quaternary geology of the Tule Springs area, Clark County, Nevada, *in* Wormington, H.M., and Ellis, D., eds., Pleistocene studies in southern Nevada: Nevada State Museum, Anthropological Papers, no. 13, p. 15–104.

Hewett, D.F., and Webber, B.N., 1931, Bedded deposits of manganese oxides near Las Vegas, Nevada: Nevada Bureau of Mines and Geology Bulletin 13, 17 p.

Hogan, J., and Bachhuber, F.W., 1981, Vegetation map of the Las Vegas SE Quadrangle: Nevada Bureau of Mines and Geology Map 3Ae.

Hunt, C.B., McKelvey, V.E., and Weise, J.H., 1942, The Three Kids manganese district, Clark County, Nevada: U.S. Geological Survey Bulletin 936-L, 319 p.

Ives, J.C., 1861, Report upon the Colorado River of the West, 1857–1858: U.S. Army, Corps of Topographical Engineers, Government Printing Office, Washington, D.C.

Jaeger, E.C., 1957, The North American deserts: Stanford University Press, Stanford, Calif., 308 p.

Jaeger, E.C., 1961, Desert wildlife: Stanford University Press, Stanford, Calif., 308 p.

Katzer, T., 1981, Flood and related debris flow hazards map of the Las Vegas SE Quadrangle: Nevada Bureau of Mines and Geology Map 3Al.

Katzer, T., Harrill, J.R., Berggren, G., and Plume, R.W., 1985, Groundwater map, Las Vegas SW Quadrangle: Nevada Bureau of Mines and Geology Map 3Bf.

Langenheim, R.L., Jr., 1963, Mississippian stratigraphy in southwestern Utah, and adjacent parts of Nevada and Arizona, *in* Guidebook to the geology of southwestern Utah: Intermountain Association of Petroleum Geologists 12th Annual Field Conference, p. 30–42.

Lanner, R.M., 1981, The piñon pine: A natural and cultural history: University of Nevada Press, Reno, 208 p.
Lanner, R.M., 1987, Trees of the Great Basin: University of Nevada Press, Reno, 215 p.
Larson, P., and Larson, L., 1997, The deserts of the Southwest, a Sierra Club naturalists guide: Sierra Club Books, San Francisco, 282 p.
Levy, M., and Christie-Blick, N., 1989, Pre-Mesozoic palinspastic reconstruction of the eastern Great Basin (western United States): Science, v. 245, p. 1454–1462.
Little, E.L., Jr., 1968, Southwestern trees—A guide to the native species of New Mexico and Arizona: U.S. Department of Agriculture, Forest Service, Agriculture Handbook, no. 9, 109 p.
Longwell, C.R., 1928, Geology of the Muddy Mountains, Nevada: U.S. Geologic Survey Bulletin 798, 151 p.
Longwell, C.R., 1949, Structure of the northern Muddy Mountains area, Nevada: Geological Society America Bulletin, v. 60, p. 923–968.
Longwell, C.R., 1974, Measure and date of movement on Las Vegas Valley shear zone, Clark County, Nevada: Geological Society of America Bulletin, v. 85, p. 985–990.
Longwell, C.R., Pampeyan, E.H., Bowyer, B., and Robert, R.J., 1979, Geology and mineral deposits of Clark County, Nevada: Nevada Bureau of Mines and Geology Bulletin 62, 218 p.
MacMahon, J.A.,1997, The National Audubon Society nature guides: Deserts: Alfred A. Knopf, New York, 638 p.
Marzolf, J.E., 1983, Changing wind and hydrologic regimes during deposition of the Navajo and Aztec Sandstones, Jurassic(?), southwestern United States, *in* Brookfield, M.E., and Ahlbrandt, T.S., eds., Eolian sediments and processes: Developments in sedimentology, v. 38, Elsevier, Amsterdam, p. 635–660.
Marzolf, J.E., 1988, Reconstruction of Late Triassic and Early and Middle Jurassic sedimentary basins—Southwestern Colorado Plateau to eastern Mojave Desert, *in* Weide, D.L., and Faber, M.L., eds., This extended land—Geological journeys in the southern Basin and Range: Field Trip Guidebook, Geological Society of America, Cordilleran Section Meeting, Las Vegas, Nev., 1988, p. 177–200.
Matti, J.C., and Bachhuber, F.W., 1985, Geologic map of the Las Vegas SW Quadrangle: Nevada Bureau of Mines and Geology Map 3Bg.
Matti, J.C., Bachhuber, F.W., Morton, D.M., and Bell, J.W., 1987, Geologic map of the Las Vegas NW Quadrangle: Nevada Bureau of Mines and Geology Map 3Dg.
Mawby, J.E., 1967, Fossil vertebrates of the Tule Springs site, Nevada, *in* Wormington, H.M., and Ellis, D., eds., Pleistocene studies in southern Nevada: Nevada State Museum, Anthropological Papers, no. 13, p. 105–128.
Maxey, G.B., and Jameson, C.H., 1948, Geology and water resources of the Las Vegas, Pahrump, and Indian Springs Valley, Clark and Nye Counties, Nevada: Nevada Department of Conservation and Natural Resources, Water Resources Bulletin 5, 128 p.
McDonnell-Canan, C., Carr, M.D., and Weide, D.L., 2000, Geologic map of the Blue Diamond NE Quadrangle, Nevada: Nevada Bureau of Mines and Geology Map 124.
McKelvey, V.E., Weise, J.H., and Johnson, V.H., 1949, Preliminary report on the bedded manganese of the Lake Mead region, Nevada and Arizona: U.S. Geological Survey Bulletin 948-D, p. 83–101.
McLane, A.R., 1974, A bibliography of Nevada caves: Center for Water Resources Research, Desert Research Institute, Reno, 99 p.
Mehringer, P.J., Jr., 1965, Late Pleistocene vegetation in the Mojave Desert of southern Nevada: Journal Arizona Academy Science, v. 3, no. 3, p. 172–188.
Mifflin, M.D., 1988, Region 5, Great Basin, *in* Back, W., Rosenshein, J.S., and Seaber, P.R., eds., Hydrogeology, v. O-2, The geology of North America: Geological Society of America, Boulder, Colo., p. 69–78.
Mifflin, M.D., and Wheat, M.M.., 1979, Pluvial lakes and estimated pluvial climate of Nevada: Nevada Bureau of Mines and Geology Bulletin 94, 57 p.
Mills, J.G., 1985, The geology and geochemistry of volcanic and plutonic rocks in the Hoover Dam 7½ minute quadrangle, Clark County, Nevada and Mojave County, Arizona [M.S. thesis]: University of Nevada, Las Vegas, 119 p.
Milne, L., and Milne, M., 1980, National Audubon Society field guide to North American insects and spiders: Alfred A. Knopf, New York, 989 p.
Mindling, A.L., 1965, An investigation of the relationship of the physical properties of fine-grained sediments to land subsidence in Las Vegas Valley, Nevada [M.S. thesis]: University of Nevada, Reno, 90 p.
Mindling, A.L., 1971, A summary of data relating to land subsidence in Las Vegas Valley: University of Nevada, Desert Research Institute Miscellaneous Report 10, 55 p.
Moehring, E.P., 1989, Resort city in the sunbelt—Las Vegas, 1930–1970: University of Nevada Press, Reno, 313 p.
Montgomery, C.W., 1990, Physical geology: W.C. Brown Publishers, Dubuque, Iowa, 555 p.
Moore, D. , 1976, Lost City: Nevada Magazine, p. 26–62.
Mozingo, H.N., 1987, Shrubs of the Great Basin: University of Nevada Press, Reno, 342 p.
Nagy, K.A., 1988, Seasonal patterns of water and energy balance in desert vertebrates: Journal of Arid Environments, v. 14, p. 201–210.
Page, W. R. and others, 2005, Map showing major structures in the Las Vegas 30' x 60' quadrangle, *in* Page, W.R., Lundstrom, S.C., Harris, A.G., Langenheim, V.E., Workman, J.B., Mahan, S.A., Paces, J.B., Dixon, G.L., Rowley, P.D., Burchfiel, B.C., Bell, J.W., and Smith, E.I., Geologic and geophysical maps of the Las Vegas 30' X 60' quadrangle, Clark and Nye Counties, Nevada, and Inyo County, California: U.S. Geological Survey SIM-2814, Sheet 1, Figure 3A. [URL:http://pubs.usgs.gov/sim/2005/2814/]
Page, W.R., Dixon, G.L., and Workman, J.B., 1998, The Blue Diamond landslide: A Tertiary breccia deposit in the Las Vegas area, Clark County, Nevada: Geological Society of America Map and Chart Series MCH083, 11 p.
Paher, S.W., 1970, Nevada ghost towns and mining camps: Howell-North Books, Berkeley, Calif., 484 p.
Papke, K.G., 1987, Gypsum deposits in Nevada: Nevada Bureau of Mines and Geology Bulletin 103, p. 14–16.
Papke, K.G., and Bell, J.W., 1978, Energy and mineral resources map of the Las Vegas SE Quadrangle: Nevada Bureau of Mines and Geology Map 3Ah.
Patterson, A., 1992, A field guide to rock art symbols of the greater Southwest: Johnson Printing Co., Boulder, Colo., 256 p.
Pavelko, M.T., Wood, D.B., and Laczniak, R.J., 1999, Las Vegas, Nevada, gambling with water in the desert, in Galloway, D., Jones, D.R., and Ingebritsen, S.E., eds., Land subsidence in the United States: U.S. Geological Survey Circular 1182, p. 49–64.
Quade, J., 1986, Late Quaternary environmental changes in the upper Las Vegas Valley, Nevada: Quaternary Research, v. 26, p. 340–357.
Quade, J., Mifflin, M.D., Pratt, W.L., McCoy, W., and Burckle, L., 1995, Fossil spring deposits in the southern Great Basin and their implications for changes in water-table levels near Yucca Mountain, Nevada, during Quaternary time: Geological Society of America Bulletin, v. 107, no. 2, p. 213–230.
Reisner, M., 1986, Cadillac desert—The American West and its disappearing water: Viking Press, New York, 582 p.
Rentz, L.H., and Smith, F.J., 1980, Plants of the Virgin Mountains: U.S. Bureau of Land Management, 51 p.
Reynolds, M.W., and Dolly, E.D., eds., 1983, Mesozoic paleogeography of the west-central United States—Rocky Mountain Paleogeography Symposium 2: Society of Economic Paleontologists and Mineralogists, Rocky Mountain Section, 573 p.
Rowland, S.M., 1986, Paleozoic, Mesozoic, and Cenozoic stratigraphy of Frenchman Mountain, Clark County, Nevada, *in* Field trip guide to the geology of southern Nevada: National Association of Geology Teachers, Far Western Section Meeting, Oct. 3–5, 1986, Las Vegas, Nev., p. 1–14.
Rowland, S.M., 1987, Paleozoic stratigraphy of Frenchman Mountain, Clark County, Nevada: Geological Society of America Centennial Field Guide—Cordilleran Section, 1987, p. 53–56.
Rowland, S.M., 1988, Southern Nevada's evolving natural landscape: Nevada Public Affairs Review, no. 1, p. 33–39.
Ryser, F. A., Jr., 1985, Birds of the Great Basin: University of Nevada Press, Reno, 603 p.
Scott, S.L., ed., 1987, Field guide to the birds of North America, second edition: National Geographic Society, Washington, D. C., 464 p.
Secor, D.T., 1962, Geology of the central Spring Mountains, Nevada [Ph.D. thesis]: Stanford University, 197 p.
Shutler, R., Jr., 1967, Archeology of Tule Springs, *in* Wormington, H.M., and Ellis, D., eds., Pleistocene studies in southern Nevada: Nevada State Museum, Anthropological Papers, no. 13, p, 297–303.
Skinner, B.J., and Porter, S.C., 1989, The dynamic Earth–An introduction to physical geology: John Wiley & Sons, New York, 540 p.
Smith, E.I., 1982, Geology and geochemistry of the volcanic rocks in the River Mountains, Clark County, Nevada and comparisons with volcanic rocks in nearby areas, *in* Frost, E.G., and Martin, D.C., eds., Mesozoic-Cenozoic tectonic evolution of the Colorado River region, California, Arizona, and Nevada: Cordilleran Publishers, San Diego, Calif., p. 41–54.
Smith, E.I., 1984, Geologic map of the Boulder Beach Quadrangle, Nevada: Nevada Bureau of Mines and Geology Map 81.
Smith, E.I., 1986, Geology of the River, Eldorado and McCullough Ranges, Clark County, Nevada, *in* Field trip guide to the geology of southern Nevada: National Association of Geology Teachers, Far West Section meeting, Oct. 3–5, 1986, Las Vegas, Nev., p. 23–35.
Smith, E.I., 1986, Road log and field guide from Henderson to Hoover Dam via Lake Mead, *in* Field trip guide to the geology of southern Nevada: National Association of Geology Teachers, Far West Section Meeting, Oct. 3–5, 1986, Las Vegas, Nev., p. 36–64.
Smith, E.I., Bridwell, H., Schmidt, C., Switzer, T., and the UNLV 1993 Winter Field Course, 1993, Late-Miocene intermediate to felsic volcanism in the McCullough Range, southern Nevada [abs.]: Journal of the Arizona-Nevada Academy of Science, v. 28, p. 45.
Smith, E.I., Feuerbach, D.L., Naumann, T.R., and Mills, J.G., 1990, Mid-Miocene volcanic and plutonic rocks in the Lake Mead area of Nevada and Arizona; production of intermediate igneous rocks in an extensional environment, *in* Anderson, J.L., ed., The nature and origin of Cordilleran magmatism: Geological Society of America Memoir 174, chapter 10, p. 169–194.
Smith, E.I., Schmidt, C.S., and Mills, J.G., 1988, Mid-Tertiary volcanoes in the Lake Mead area of southern Nevada and northwestern Arizona, *in* Weide D.L., and Faber, M.L., eds., This extended land-Geological journeys in southern Basin and Range: Field Trip Guidebook, Geological Society of America, Cordilleran Section Meeting, Las Vegas, Nev., 1988, p. 107–122.
Sowell, J. 2001, Desert ecology, an introduction to life in the arid southwest: University of Utah Press, Salt Lake City, 193 p.
Sowers, J.M., Amundson, R.G., Chadwick, O.A., Harden, J.W., Jull, A.J.T., Ku, T.L., McFadden, L.D., Reheis, M.C., Taylor, E.M., and Szabo, B.J, 1988, Geomorphology and pedology on the Kyle Canyon alluvial fan, southern Nevada, *in* Weide, D.L., and Faber, M.L., eds., This extended land—Geological journeys in the southern Basin and Range: Field Trip Guidebook, Geological Society of America, Cordilleran Section Meeting, Las Vegas, Nev., 1988, p. 137–157.
Spaulding, W.G., 1985, Vegetation and climates of the last 45,000 years in the vicinity of the Nevada Test Site, south-central Nevada: U.S. Geological Survey Professional Paper 1329, 83 p.
Stevens, J.E., 1988, Hoover Dam—An American adventure: University of Oklahoma Press, Norman, Okla., 326 p.
Stewart, J.H., 1980, Geology of Nevada: Nevada Bureau of Mines and Geology Special Publication 4, 136 p.
Taylor, E.M., 1989, Late Quaternary paleoclimate studies—Geologic problems and questions, *in* Late Cenozoic evolution of the southern Great Basin: Nevada Bureau of Mines and Geology Open-File Report 89-1, p. 153.
Taylor, R. J., 1992, Sagebrush country: a wildlife sanctuary: Mountain Press Publishing Co., Missoula, Montana, 211 p.
Terres, J.K., 1980, The Audubon Society encyclopedia of North American birds: Alfred A. Knopf, Inc., New York, 1109 p.
U.S. Geological Survey, 1976, Field trip to Nevada Test Site: U.S. Geological Survey Open-File Report 76-313, 64 p.
U.S. Soil Conservation Service, 1982, Soils map of the Las Vegas SE Quadrangle: Nevada Bureau of Mines and Geology Map 3Ad.
Van Devender, T.R., 1977, Holocene woodlands in the southwestern deserts: Science, v. 198, p. 189–192.
Van Devender, T.R., and Spaulding, W.G., 1979, Development of vegetation and climate in the southwestern United States: Science, v. 204, p. 701–710.
Walker, T.R., and Honea, R.M., 1969, Iron content of modern deposits in the Sonoran Desert—A contribution to the origin of Red Beds: Geological Society of America Bulletin, v. 80, p. 535–544.
Wallin, E.T., Duebendorfer, E.M., and Smith, E.I., 1993, Tertiary stratigraphy of the Lake Mead region, *in* Sherrod, D.R., and Nielson, J.E., eds., Tertiary stratigraphy of highly extended terranes, California, Arizona, and Nevada: U.S. Geological Survey Bulletin 2053, p. 33–35.
Weber, M.E., and Smith, E.I., 1987, Structural and geochemical constraints on the reassembly of mid-Tertiary volcanoes in the Lake Mead area of southern Nevada: Geology, v. 15, p. 553–556.
Wernicke, B., Axen, G.J., and Snow, J.K., 1988, Basin and Range extensional tectonics at the latitude of Las Vegas, Nevada: Geological Society of America Bulletin, v. 100, p. 1738–1757.
Wernicke, B., Guth, P.L., and Axen, G.J., 1984, Tertiary extensional tectonics in the Sevier thrust belt of southern Nevada, *in* Lintz, J., Jr., ed., Western geological excursions: Geological Society of America 1984 Annual Meeting, Reno, Guidebook, v. 4, p. 473–510.
Wheeler, S.S., 1982, The Nevada desert: The Caxton Printers, Ltd., Caldwell, Idaho, 168 p.
Whitney, S., 1997, The National Audubon Society nature guides: western forests: Alfred A. Knopf Inc., New York, 670 p.
Wright, F., 1981, Clark County—The changing face of southern Nevada: Nevada Historical Society, Las Vegas, 38 p.
Wright, F., 1984, The pioneering adventure in Nevada: Nevada Historical Society, Las Vegas, 29 p.
Zdon, A., and Kepper, J., 1991, Las Vegas to Hoover Dam, and the Frenchman Mountain area, *in* Geology of the Las Vegas region: American Association of Professional Geologists, Nevada Section, 1991 Field Trip, p. 1–10.

Glossary

adit A horizontal or nearly horizontal passage from the surface into a mine.

agglomerate A pyroclastic rock containing a predominance of rounded or subangular fragments mostly greater than 1 inch in diameter.

alkali flat A level area or plain in an arid or semiarid region, usually low-lying with poor drainage, encrusted with alkali salts that became concentrated by evaporation; a salt flat. See also: playa.

alluvial fan A fan-shaped deposit of alluvium typically built where a stream leaves a steep mountain valley and runs out onto a level plain.

alluvial terrace A stream terrace composed of unconsolidated alluvium (including gravel), produced by renewed downcutting of the flood plain or valley floor by a rejuvenated stream.

alluvium A general term for clay, silt, sand, gravel, or similar unconsolidated material, deposited during comparatively recent geologic time by a stream or other body of running water.

altered A change in the mineral composition of a rock that is usually brought about by the action of hydrothermal solutions or by weathering. Hard, rock-forming minerals such as feldspar typically "alter" to softer minerals such as clay and micas.

alunite A mineral, hydrated potassium aluminum sulfate, commonly found as soft white, gray, or pink masses in hydrothermally altered rocks that originally contained abundant feldspar.

andesite A fine-grained volcanic rock that solidifies from molten lava at the Earth's surface. It is intermediate in composition between basalt and rhyolite, and ranges in color from dark gray-green to lighter gray, red or brown.

anhydrite An evaporite mineral (anhydrous calcium sulfate) commonly found with gypsum in massive beds. It is translucent with glassy to pearly luster, and is generally white or gray.

anhydrous A substance, e.g. magma or a mineral, that is completely or essentially without water. An anhydrous mineral contains no water in chemical combination.

anticline An arching fold in which the limbs dip away from the axis. The oldest rocks occur at the center of the fold, with progressively younger rocks found outward from the core.

aquifer A body of rock that is sufficiently permeable to allow groundwater to flow through it and to yield water to wells and springs.

ash, volcanic Fine pyroclastic material produced by the explosive emission of hot, gas-charged lava from a volcanic crater or fissure. The material cools as it falls to the ground, producing fragments usually light gray and mostly less than 1/4 inch in diameter.

arrastre A historical apparatus for grinding and mixing ores by means of a heavy stone dragged around upon a circular bed. The arrastre was chiefly used for ores containing free gold.

asthenosphere The weak, plastic, partly molten layer of the upper mantle directly below the lithosphere. It lies 60 to 220 miles below the Earth's surface.

badlands A region nearly devoid of vegetation where erosion has cut into soft, easily erodible rock forming an intricate maze of narrow ravines and sharp crests and pinnacles.

bajada A series of coalescing alluvial fans along the base of a mountain range.

barite Barium sulfate.

basal conglomerate A conglomerate that forms the bottom stratigraphic unit of a sedimentary series and that rests on a surface of erosion, thereby marking an unconformity.

basalt A fine-grained igneous rock that solidifies from molten lava at the Earth's surface. It is usually black, due to the predominance of dark-colored minerals.

basaltic andesite An igneous rock intermediate in composition between andesite and basalt. Its color is commonly darker than andesite and lighter than basalt.

basement The undifferentiated complex of rocks (generally igneous or metamorphic) that underlie surface rocks.

Basin and Range province A physiographic region in the western United States that consists of fault-block mountains and intervening sediment-filled basins. The province lies between the Sierra Nevada on the west, the Columbia Plateau and Snake River Plain on the north, and the Colorado Plateau on the east. On the south, the province extends through southern Arizona and into northern Mexico.

basin-and-range extension Extension, or pulling apart, of the crust of the Earth in the western United States has produced the present physiography of generally north-trending mountain ranges separated by intervening sediment-filled basins.

batholith A large body of intrusive igneous (plutonic) rock, often produced by multiple intrusions.

bedding Depositional layers or planes dividing sedimentary rocks of the same or different lithology.

bedrock Any solid rock exposed at the Earth's surface or covered by unconsolidated material.

benchmark A surveyor's mark made on some stationary object of previously determined position and elevation, and used as a reference point in surveys.

biotite A common rock-forming mineral and a member of the mica group of minerals. It ranges from dark brown to green and exhibits perfect basal cleavage, that is, it will peel into thin, transparent layers along one plane.

bleaching A lightening of the original color of rock; a surface effect caused by long exposure to weathering or to (generally) hot, circulating solutions, that have altered the original chemical composition of the minerals forming the rocks.

borates Oxide compounds of the element boron; includes the minerals borax, colemanite, and ulexite.

borax Hydrous sodium borate.

brachiopod A marine invertebrate belonging to the phylum Brachiopoda, characterized by two bilaterally symmetrical valves. Range, Early Cambrian to present.

breccia A coarse-grained rock composed of angular fragments of broken rock in a finer-grained matrix; may or may not be cemented.

butte A conspicuous isolated hill or small mountain with very steep sides, a small mesa.

calcite Calcium carbonate, the principal constituent of limestone.

calcrete A term for a conglomerate consisting of sand and gravel cemented into a hard mass by calcium carbonate precipitated from infiltrating groundwater.

caldera A large, bowl-shaped volcanic depression with a diameter many times greater than the included volcanic vent or vents. It may be formed by explosion or collapse.

caliche A solid, almost impervious accumulation of whitish calcium carbonate-rich material commonly found in layers on or near the surface of soils in arid regions.

cap rock A hard rock layer capping a softer rock unit (such as basalt capping a soft sandstone).

carbonate rock A rock consisting mainly of carbonate minerals, such as limestone or dolomite.

chert A sedimentary rock composed of extremely fine-grained quartz, formed by organic or inorganic precipitation or by replacement. Flint and jasper are varieties of chert.

chlorination A process used to recover gold from base-metal ore in which the ore is first roasted to oxidize the base metals, then saturated with chlorine gas, and finally treated with water, which removes the soluble chloride of gold, to be subsequently precipitated and melted into bars.

cinder cone A conical hill formed by the accumulation of volcanic ash or cinders around a vent.

cinder, volcanic Uncemented, glassy, vesicular rock ejected from a volcanic vent.

claim post A post or rock monument used to mark the boundary of a mining claim.

clast A piece of broken rock or an individual constituent of sedimentary rock produced by the physical disintegration of a larger rock mass.

clay An extremely fine-grained, natural sediment or soft rock composed of particles less than 4 microns.

colemanite A mineral (hydrous calcium borate) that is colorless to white, transparent to translucent, with a glassy luster.

compressional deformation Deformation of rocks due to compressive (squeezing) forces within the Earth's crust.

concretion A hard, nodular body formed within and enclosed by softer sedimentary rock.

conglomerate A sedimentary rock consisting of boulders and cobbles set in a finer-grained matrix.

construction aggregate A variety of materials including sand, gravel, crushed stone, and cinders that are used to provide bulk and strength in portland cement concrete, asphalt concrete, fill, road base and loose road surfacing, railroad ballast, concrete block, and stucco.

contact The place or surface where two different kinds of rocks come together.

continental crust The portion of the Earth's crust that underlies the continents.

continental plate A thick portion of the Earth's crust that comprises a continent.

corrasion Erosion of rock and soil by the abrasive action of particles set in motion by running water, wind, glaciers, or gravity.

cottonball borax See ulexite.

cross-bedding A sequence of beds inclined at an angle to the main bedding planes in granular sediments.

crust The outermost compositional shell of the Earth, 6 to 25 miles thick.

crustal extension Separation, or pulling apart, of the Earth's crust due to forces in the lower crust and mantle, such as convection currents, which are believed to "drive" the continental and oceanic plates of the Earth's crust.

cyanidation A process involving the use of cyanide (potassium or sodium cyanide), in the extraction and recovery of gold and silver from ore.

dacite A light-colored volcanic rock about midway between andesite and rhyolite in composition and appearance.
debris flow The downslope movement of a mass of unconsolidated and unsorted rock with associated water and mud.
denudation fault See detachment fault.
desert pavement A natural residual concentration of wind-polished, closely packed pebbles, boulders, and other rock fragments, mantling a desert surface where wind action and sheetwash have removed all smaller particles, usually protecting the underlying finer-grained material from further erosion.
desert varnish See rock varnish.
detachment fault A large-scale low-angle normal or thrust fault.
detritus A collective term for loose rock and mineral material that is worn off or removed by mechanical means, as by disintegration or abrasion; esp. fragmental material, such as sand, silt, and clay, derived from older rocks and moved from its place of origin.
differential weathering Weathering that occurs at different rates or intensity as a result of variations in the composition and structure of rocks.
dike A tabular sheet of intrusive igneous rock that cuts across the structure of the intruded rock or cuts massive rock.
dip The angle in degrees between the horizontal and an inclined geologic structure, such as a bedding plane or a fault.
dip slope A slope of the land surface that conforms approximately to the dip of the underlying rocks.
dissected Cut by erosion into hills and valleys.
dissolution The chemical weathering process whereby minerals and rock material pass directly into solution.
dolomite a) a mineral, calcium magnesium carbonate. It has a glassy or pearly luster and is usually some shade of gray, tan, or pink. b) a carbonate sedimentary rock of which more than 50 percent by weight consists of the mineral dolomite. Dolomite occurs in crystalline and noncrystalline forms, is clearly associated and often interbedded with limestone, and usually represents a postdepositional replacement of limestone.
dolostone See dolomite, the carbonate sedimentary rock.
dome A large body of plutonic or volcanic rock that, in outcrop, resembles the dome or cupola of a building.
drag The bending of strata on either side of a fault, caused by the friction of the moving blocks along the fault.
drift A horizontal passage in an underground mine. A drift follows along a vein, as distinguished from a crosscut, which crosses a vein.
earthquake A sudden motion or trembling in the Earth's crust caused by the abrupt release of slowly accumulated strain.
ephemeral stream A stream that flows occasionally in direct response to precipitation or snowmelt.
erosion The group of related processes by which rock is broken down physically and chemically and the products removed from any part of the earth's surface. It includes the processes of weathering, solution, corrasion, and transportation.
escarpment A long, generally continuous cliff or steep face at the edge of a region of high local relief.
evaporite A mineral deposit formed by the evaporation of water in a restricted basin; also, the minerals of such deposits.
extension In geology, horizontal expansion or pulling apart of the Earth's crust.
extensional deformation The deformation rocks undergo when subjected to extension, generally consisting of low- to high-angle normal faulting.
extrusive Igneous rocks formed from magmas or magmatic materials that are poured out or ejected onto the Earth's surface by volcanic activity.
fan See alluvial fan.
fault A fracture or planar break in rock (which may many miles long) along which there is movement of one side relative to the other.
fault breccia Crushed and broken rock adjacent to a fault formed by the mechanical breakup of rocks during displacement along the fault.
fault gouge Finely abraded material occurring between the walls of a fault, the result of grinding during movement.
fault scarp The steep slope or cliff formed by a fault. Most fault scarps have been modified by erosion since the faulting.
fault surface The surface along which movement has taken place along a fault.
fault trace The line of intersection of a fault plane with the Earth's surface.
fault zone A fault, instead of being a single clean fracture, may be a zone as much as hundreds or even thousands of feet wide; the fault zone consists of numerous interlacing small faults or a complex zone of fault gouge or breccia.
feldspar A silicate mineral common in many rocks and making up 60 percent of the Earth's crust.
fissure A extensional crack, break, or fracture in the rocks.
flash flood A local, sudden flood of water through a stream channel, generally of relatively great volume and short duration after intense local precipitation.
floodplain That portion of a river valley that is covered with water when the river overflows its banks.
fold A curve or bend of a planar structure such as rock strata, bedding planes, foliation, or cleavage.
footwall The lower side of a fault plane, vein, lode, or bed of ore. So named because miners in underground developments along a vein stood on the "foot" wall.
formation A distinctive, mappable rock unit representing deposition under a uniform set of conditions and during a limited period of time.
fossil Any remains, trace, or imprint of a plant or animal that has been preserved in the Earth's crust since some past geologic or prehistoric time; loosely, any evidence of past life.
garnet A group of aluminum-silicate minerals containing variable amounts of calcium, magnesium, iron, manganese, and chrome. Garnet is a brittle and transparent to subtransparent mineral, having a vitreous luster, no cleavage, and a variety of colors, dark red being the most common.
gneiss A coarse-grained metamorphic rock exhibiting a banded texture (foliation). Bands rich in granular minerals alternate with bands of flaky or elongate minerals.
granite A light-colored coarse-grained, igneous (plutonic) rock containing the minerals quartz and orthoclase feldspar (light-colored minerals), with lesser amounts of plagioclase feldspar, mica, and hornblende (dark-colored minerals).
granitic rock Granite or a close relative, such as granodiorite or quartz monzonite.
granodiorite A coarse-grained, plutonic rock similar to granite, but with almost equal amounts of dark and light-colored minerals giving this rock a salt-and-pepper appearance.
gravel An unconsolidated, natural accumulation of rounded rock fragments resulting from erosion, consisting predominantly of particles larger than sand, such as boulders, cobbles, pebbles, granules, or any combination of these fragments.
groundwater Loosely, all subsurface water as distinct from surface water.
gypsum A mineral, hydrous calcium sulfate. Gypsum is soft and white or colorless when pure, and frequently forms thick, extensive beds interstratified with limestone, shale, and clay. It occurs massive (alabaster), fibrous (satin spar), or in crystals (selenite).
halite Native salt occurring in massive, granular, compact, or cubic-crystalline forms.
hanging wall The rock on the upper side of a fault, mineral vein, or mineral deposit.
headward erosion The lengthening and cutting upstream of a valley or gully above the original source of its stream. It is accomplished by erosion due to rainwash, gullying, spring sapping, and the slumping of material into the head of the growing valley.
hematite A mineral (iron oxide) that is reddish brown to black or silver gray. An important ore of iron.
hogback ridge Any ridge with a sharp summit and steep slopes of nearly equal inclination on both flanks, and resembling in outline the back of a hog. The term is usually restricted to ridges carved from beds dipping at angles greater than 20 degrees
hoodoo A tall, conical column of unconsolidated to semiconsolidated material such as. clay, till, or landslide debris, produced by differential erosion in a region of sporadic heavy rainfall (as in a badland or a desert wash), and usually capped by a flat, hard boulder that shielded the underlying softer material from erosion.
hornblende A common rock-forming dark silicate mineral.
hourglass canyon A rugged canyon, common in desert mountains, wide at the top with a constricted mouth. The widening cliffs above the constriction and the spreading alluvial fan below give the impression of an hourglass.
hydrous mineral A mineral compound containing water.
igneous rock Rock formed by the cooling and consolidation of magma.
impermeable The condition of a rock, sediment, or soil that does not allow the flow of water.
industrial mineral Any rock or mineral, or other naturally occurring substance of economic value, exclusive of metallic ores, mineral fuels, and gemstones.
interbedded Alternating layers of different materials in a section of bedded rocks.
intermontane basin A basin situated between or surrounded by mountains, mountain ranges, or mountainous regions.
intrusive Plutonic. Magma that penetrated into or between other rocks and solidified before reaching the Earth's surface.
joint A fracture in rock along which no appreciable displacement has occurred.
laccolith A igneous intrusive that has moved into and formed along bedding planes in older rocks. Laccoliths have a known or assumed flat floor and a postulated dikelike feeder.

landform Any feature of the Earth's surface having a characteristic shape as the product of natural processes. Examples are continents, ocean basins, mountains, alluvial fans, sand dunes, and valleys.

landslide A general term covering the failure and rapid movement of rock masses downslope on the earth's surface.

lava A general term for a molten extrusive; also, for the rock that is solidified from it.

lava flow Magma that flows out at the Earth's surface.

lava shield A shield volcano. A volcano with a low, flat, broad shape, formed by the buildup of many thin lava flows.

left-lateral fault A strike-slip fault in which relative motion is such that to an observer looking directly at (perpendicular to) the fault, the motion of the block on the opposite side of the fault is to the left.

life zone A zone defined by elevation and latitude that hosts a characteristic assemblage of plants.

limestone A sedimentary rock consisting chiefly of calcium carbonate, mainly in the form of the mineral calcite.

lithosphere The rigid, outermost layer of the Earth, 30 to 60 miles thick, encompassing the crust and upper mantle.

low-angle normal fault A gently inclined, dip-slip fault in which the hanging wall has moved downward relative to the footwall.

lower plate The footwall or lower side of an inclined fault.

magma Molten rock that is generated when temperatures rise and melting occurs in the mantle or crust of the Earth. Igneous rocks are formed when magma cools and consolidates.

magma chamber A large reservoir in the Earth's crust occupied by magma.

magmatic Of, pertaining to, or derived from magma.

magnesite A mineral (magnesium carbonate). Magnesite is generally found as white to grayish, yellow, or brown earthy masses or irregular veins resulting from the alteration of dolomite rocks, or of rocks rich in magnesium silicates.

manganite A mineral (hydrous manganese oxide). It is characteristically opaque, steel-gray to black, and has long, prismatic crystals.

mesa A tableland; a flat-topped mountain or other elevated landform bounded on at least one side by a steep cliff.

metamorphic rock Rock whose original textures or mineral components, or both, have been transformed to new textures and components as a result of high temperature, high pressure, or both.

microcline A feldspar mineral (calcium aluminum silicate) that exhibits a glassy luster, is transparent to translucent, and varies from white to pale yellow or pink.

mineral A naturally occurring, inorganic, solid element or compound, with a definite composition or compositional range and a regular internal crystal structure.

mining claim That portion of the public lands that a miner acquires for mining purposes and holds in accordance with the mining laws.

monzonite A granular plutonic rock resembling granite that contains approximately equal amounts of orthoclase and plagioclase (feldspars). Quartz is usually present in very minor quantities (less than 2 percent of the volume of the rock). Biotite and hornblende are usually present, along with lesser amounts of apatite, zircon, and sphene.

mud crack An irregular fracture in a crudely polygonal pattern, formed by the shrinkage of clay, silt, or mud in the course of drying.

mudflat A relatively level area of fine silt along a shore (as in a sheltered estuary) or around an island, alternately covered and uncovered by the tide, or covered by shallow water; a muddy tidal flat barren of vegetation.

mudflow A general term for a mass-movement landform and a process characterized by a flowing mass of predominantly fine-grained material mixed with water (mud).

mudstone A rock that includes clay, silt, siltstone, claystone, shale, and/or argillite. The term is used when precise rock identification is in doubt.

neotocite A mineral (hydrous manganese-iron silicate).

nonmetallic mineral See industrial mineral.

normal fault An inclined fault along which the hanging wall has moved downward relative to the footwall.

oceanic crust The Earth's crust that underlies the ocean basins.

oceanic plate A tectonic plate of the Earth's crust that underlies an ocean.

ore The naturally occurring material from which a mineral or minerals of economic value can be extracted at a reasonable profit.

orthoclase A mineral (potassium aluminum silicate); a member of the feldspar group and a common mineral in granitic rocks. Its luster is glassy, and it ranges from colorless to white, gray, or flesh-red.

outcrop That part of a geologic formation that appears at the surface of the Earth.

oxidation The process of combining with oxygen; a common alteration process

paleohydrologic Ancient hydrologic features preserved in rock.

pediment An eroded bedrock surface that slopes away from the base of mountains in arid regions and is thinly or discontinuously covered by alluvium.

pediment gravel The gravel derived from the mountains upslope that form the cover on a pediment.

pegmatite A very coarse-grained igneous rock, generally containing the same minerals as granite, that commonly occurs in dikes or veins.

permeable Having pores that permit fluids to pass through.

petrified wood Fossilized wood, formed when wood is buried and replaced by an equal volume of mineral matter.

petroglyph Writings or drawings chipped or scraped into the surface of rock faces. Usually the markings are made in the veneer of dark rock varnish on the face of lighter-colored rock for good contrast.

phenocryst One of the larger, isolated crystals in a porphyritic igneous rock.

phreatophyte A plant that obtains its water supply from the zone of saturation or through the capillary fringe and is characterized by a deep root system.

plate tectonics The processes or mechanisms by which the Earth's lithosphere (crust and upper mantle) is broken up into a series of rigid plates that move over the asthenosphere (weak layer in upper mantle).

playa The flat, vegetation-free, lowermost area of a desert basin, where water gathers after a rain and evaporates.

playa lake A shallow, intermittent lake in an arid or semiarid region, covering or occupying a playa in the wet season but drying up in summer; an ephemeral lake that upon evaporation leaves a playa.

pluton Any body of igneous rock that has formed beneath the surface of the earth by solidification of magma, regardless of shape or size.

plutonic rock Igneous rock that has solidified at relatively great depth below the Earth's surface (the rock that makes up a pluton).

porphyritic An igneous rock that has coarse crystals in a finely crystalline or glassy groundmass.

potash Potassium carbonate.

psilomelane A mineral (manganese oxide). It is black with an opaque, sub-metallic luster, and rarely exhibits an apparent crystal structure. It readily soils the hands.

pumice A natural glassy froth made by gases escaping through a viscous magma.

pyroclastic A general term applied to volcanic materials that have been explosively or aerially ejected from a volcanic vent. Also, a general term for the class of volcanic rocks made up of these materials.

pyrolusite A mineral and the principal ore of manganese, (manganese dioxide). It has an opaque, metallic luster; is iron-black; and fractures into splinters. It is found as radiating fibers or columns, coatings, dendritic shapes, or granular masses. It readily soils the hands.

quarry An open or surface working, usually for the extraction of building stone or other industrial minerals.

quartz An important rock-forming mineral (silicon dioxide). Quartz is second only to feldspar as the most common rock-forming mineral, and it forms the major proportion of most sands. It has a widespread distribution in igneous (especially granitic), metamorphic, and sedimentary rocks. Quartz has a glassy luster and fracture, and lacks cleavage.

quartz monzonite A granular plutonic rock that resembles and is related to granite. Its major constituents are orthoclase, plagioclase and quartz, with minor quantities of biotite, hornblende, apatite, and zircon.

rapakivi A rock texture originally described in Finnish granites. In typical specimens flesh-colored orthoclase feldspar occurs as large, rounded crystals up to an inch in diameter. These feldspar crystals are embedded in a matrix that has normal granitic texture, but consists chiefly of quartz and colored minerals.

reverse fault An inclined fault along which the hanging-wall block has moved upward relative to the footwall.

rhyolite A fine-grained volcanic rock with the same composition as granite.

rhyolitic See rhyolite

right-lateral fault A strike-slip fault in which relative motion is such that to an observer looking directly at (perpendicular to) the fault, the motion of the block on the opposite side of the fault is to the right.

riparian The association of life, specifically plant life, living and growing on the bank of a spring, stream, river, lake, or other water body.

ripple mark An undulating surface of alternating subparallel small-scale ridges and hollows produced on a sediment on land by wind action and under water by currents or by the agitation of water in wave action.

rock Any naturally formed, solid aggregate of one or more minerals.

rock varnish A thin dark shiny film or coating, composed of iron oxide accompanied by traces of manganese oxide and silica, formed on the surfaces of pebbles, boulders, and other rock fragments after long exposure, as well as on ledges and other rock outcrops. Also called desert varnish.

salt dome A nearly equidimensional structure with a central salt plug which has risen by forcing its way through the enclosing sediments from a mother salt bed below. Most plugs have nearly vertical walls, but some overhang. The resulting dome can be 0.5 to 1 mile or more in diameter.

San Andreas fault A major strike-slip fault zone extending from the north-central coast of California through the Gulf of California.

sand A rock fragment or particle 0.05 to 2 mm in size. The material is most commonly composed of quartz resulting from rock disintegration.

sandstone A clastic sedimentary rock made up of sand-sized particles.

scarp See fault scarp.

schist A metamorphic, crystalline rock that can be readily split into thin flakes or slabs.

sea-floor spreading The process that increases oceanic crust by convective upwelling of magma along the mid-oceanic ridges. The new material created moves away from the ridges at a rate of one to ten centimeters per year (see plate tectonics).

sediment Material that weathers from rocks and is transported or deposited by air, water, or ice, or that accumulates by other natural agents, such as chemical precipitation from solution or secretion by organisms, and that forms in layers on the Earth's surface at ordinary temperatures in a loose, unconsolidated form; e.g. sand, gravel, silt, mud, alluvium.

sedimentary rock A consolidated accumulation of rock and mineral grains and organic matter or a rock that has been formed by chemical or organic precipitation.

selenite The clear, colorless variety of gypsum, occurring in distinct, transparent crystals or in large crystalline masses that easily cleave.

shaft A mine working, extending from surface down; can be vertical or inclined. Used to move men and equipment in and out of the mine and to move ore out of the mine.

shale A very fine-grained, laminated, clastic sedimentary rock made up of clay-sized particles, none of which are larger than 4 microns. Shale tends to break along parallel planes.

shear zone A fault zone in which movement has occurred on a large scale so that the rock is crushed and brecciated.

sheared Rock affected (usually broken and crushed) by a shear zone.

shield volcano A volcano with a low, flat, broad shape, formed by the build-up of many thin, generally basaltic, lava flows.

silica Silicon dioxide.

silt A clastic (made of individual particles) sediment in which most of the particles are between 0.002 and 0.05 mm in diameter.

siltstone A very fine-grained, consolidated, clastic sedimentary rock composed predominantly of silt-size particles.

slickensides Striated or highly polished surfaces on hard rocks abraded during movement along a fault.

specular hematite A variety of the mineral iron oxide, hematite, that occurs in tabular crystals having a gray, foil-like, metallic luster.

spherulite A small spherical aggregation of one or more minerals that is formed by the radial growth of crystals about a common center or inclusion. Such structures are especially common in the glassy groundmass of some lava flows and in obsidian.

spreading ridge See sea-floor spreading.

stamp mill A historical apparatus in which rock was crushed by descending iron pestles (stamps) generally grouped in units (batteries) of five per mortar. Amalgamation (collection with mercury) was usually combined with stamp milling to recover gold and silver from the crushed rock.

stock A body of plutonic rock that is similar to, but smaller than, a batholith.

stope An excavation in an underground from which ore has been extracted.

strata Layers or tabular beds of sedimentary rock that consist of approximately the same kind of material throughout, and that are distinct from the layers above and below.

stratiform Having the form of a layer, bed, or stratum; consisting of roughly parallel bands or sheets.

stratification The layered arrangement of sediments, sedimentary rocks, or extrusive igneous rocks.

stratovolcano A volcano that emits both fragmental material and viscous lava, and that builds up steep conical cones. Also called a composite volcano.

striated Solid rock that has had parallel grooves cut into its surface by fault movement or by movement of glacial ice.

strike The direction, measured as an angle from true north, of a horizontal line in the plane of an inclined rock unit, joint, fault or other structural plane. It is perpendicular to the dip.

strike-slip fault A high-angle fault along which displacement has been horizontal. See right-lateral and left-lateral strike-slip fault.

subduction zone A convergent plate boundary at which a slab of oceanic lithosphere sinks beneath another plate (continental or oceanic) and carried down into the mantle of the Earth.

subsidence The sudden sinking or gradual downward settling of the Earth's surface with little or no horizontal motion. The movement is not restricted in rate, magnitude, or area involved. Subsidence may be caused by natural geologic processes, such as solution, thawing, compaction, slow crustal warping, or withdrawal of fluid lava from beneath a solid crust; or by man's activity, such as subsurface mining or the pumping of oil or ground water.

syncline A downfold with a troughlike form. The youngest rock will be found in the center of the fold, with progressively older rocks exposed away from the center.

tailings The rock waste left after most of the valuable mineral has been extracted from an ore; the part rejected in milling an ore.

talus Rock fragments of any size or shape (usually coarse and angular) derived from and lying at the base of a cliff or very steep, rocky slope. Also, the outward sloping and accumulated heap or mass of such loose broken rock, considered as a unit, and formed chiefly by gravitational falling, rolling, or sliding.

talus cone A cone-shaped or apron-like landform at the base of a cliff, consisting of poorly sorted rock debris that has accumulated episodically by rockfall or alluvial wash.

tear fault A strike-slip fault that trends transverse to the strike of the deformed rocks. Usually restricted to strike-slip faults that are confined to upper plate of either a major thrust fault or a major low-angle normal (detachment) fault.

thrust fault A low-angle reverse fault with the fault plane dipping less than 45 degrees.

thrust sheet The body of rock above a large-scale thrust fault whose surface is horizontal or very gently dipping.

topographic slope The slope of the local land surface or relief.

transpiration The process by which water absorbed by plants, usually through the roots, is evaporated into the atmosphere from the plant surface.

trilobite A marine arthropod that lived during the Paleozoic Era; characterized by its three-part (tri-lobed) body.

tuff A rock formed of compacted volcanic fragments. A general term for all consolidated pyroclastic rocks.

tunnel Strictly speaking, a passage in a mine that is open to the surface at both ends. It is often used incorrectly as a synonym for adit, which has only one opening to surface; or drift, which is driven underground within a mine and has no openings to surface.

unconformity A substantial gap or break in the geologic record. A surface within a sedimentary sequence that records a period of nondeposition or erosion.

upper plate The hanging wall or upper side of an inclined fault.

ulexite A white borate mineral (hydrous sodium-calcium borate). It forms rounded masses of extremely fine needle-like crystals and is usually associated with borax in saline crusts on alkali flats in arid regions.

vein A mineral-filled fault or fracture.

vent The conduit and orifice through which volcanic materials (lava, gas, and water vapor) reach the Earth's surface.

vesicular basalt basalt lava, characterized by abundant vesicles formed as a result of the expansion of gases during the fluid stage of the lava.

volcanic ash See ash, volcanic

volcanic dome See dome

volcanic neck The solidified material filling a vent or pipe of an inactive volcano. This hard igneous rock may resist erosion better than the mountain mass originally encompassing it and eventually stand alone as a column, tower, or crag.

volcanic plug Necks consisting of a monolithic mass of solidified volcanic rock.

volcanic rock Igneous rocks derived from magma or magmatic materials that are poured out or ejected (extruded) at or very near the Earth's surface.

volcano A vent from which magma, gas, and ash are erupted; also, the usually conical structure built by such eruptions.

wad An impure mixture of manganese oxide and other oxides. It contains 10 to 20 percent water, is generally soft and black, and readily soils the hands.

wash A shallow streambed with steep sides cut into unconsolidated sediments. This kind of streambed usually carries water only after brief, local precipitation.

water table The upper limit of the portion of the ground wholly saturated with water.

wetland A general term for a group of wet habitats. It includes areas that are permanently wet or intermittently covered by water.

Index

GPS Coordinates for Trips

Trip 1

GPS Point	Cumulative mileage	Position	Comments
		Lat/Long ddd°mm.mmm'	Datum: WGS 84
1	0.0	N36 02.517 W115 11.275	Dean Martin Drive (formerly Industrial Road)
2	2.5	N36 01.530 W115 13.593	Arden railroad crossing
3	4.2	N36 01.223 W115 15.407	Exploration Peak to left
4	6.0	N36 01.222 W115 17.397	Mile point 6.0
5			*Point not used*
6			*Point not used*
7	8.1	N36 01.667 W115 19.512	Limestone ridge to the right at 3:30
8			*Point not used*
9	10.0	N36 02.112 W115 21.525	Intersection with State Route 159 on right
10	11.8	N36 02.940 W115 23.065	Intersection with road to Blue Diamond plant of BPB
11	12.2	N36 02.985 W115 23.498	Narrows, Kaibab limestone on ridge
12	12.5	N36 02.947 W115 23.792	Sign, entering Red Rock Canyon National Conservation Area, on right
13	13.1	N36 03.052 W115 24.330	Road into town of Blue Diamond on left (Castalia St.)
14	13.7	N36 03.235 W115 24.935	View ahead of Wilson Cliffs and First Creek Canyon
15			*Point not used*
16	15.0	N36 03.953 W115 26.115	Bonnie Springs Ranch turn on left
17	15.8	N36 04.433 W115 26.615	Spring Mountain Ranch State Park turn on left
18	16.4	N36 04.887 W115 26.882	First Creek Trailhead on left
19	17.0	N36 05.375 W115 26.937	Oak Creek turn on left
20	18.9	N36 07.008 W115 26.698	Red Rock Overlook and display on left
21	19.2	N36 07.142 W115 26.397	Private road to Blue Diamond Mine on right
22	20.7	N36 07.905 W115 25.285	Turn into entrance road, Red Rock Canyon National Conservation Area
	0.0	*(odometer reset)*	
23		N36 07.977 W115 25.445	Begin Red Rock scenic drive
24	1.0	N36 08.737 W115 25.833	Calico Hills No. 1 Overlook on right
25	1.5	N36 09.117 W115 26.245	Calico Hills No. 2 Overlook on right
26	2.4	N36 09.627 W115 26.932	Sandstone Quarry road on right
27	4.7	N36 10.253 W115 27.958	High Point Overlook on left
28	5.8	N36 09.965 W115 28.510	White Rock Spring turn on right
29	8.4	N36 09.363 W115 29.407	Willow Spring Picnic area turn on right
30	10.1	N36 09.022 W115 29.053	Ice Box Canyon Trailhead on right
31	10.7	N36 08.675 W115 28.607	Crossing Red Rock Wash
32	11.0	N36 08.527 W115 28.395	Red Rock Wash Overlook
33	12.7	N36 07.765 W115 28.397	Pine Creek Canyon Overlook and Trailhead on right
34	14.0	N36 07.015 W115 27.402	Oak Creek Trailhead on right
35	14.6	N36 06.672 W115 26.973	End of Red Rock scenic drive, intersection SR 159
36	18.3	N36 08.498 W115 23.983	Calico Basin-Red Rock Spring road on left
37	18.8	N36 08.630 W115 23.493	Road to 13-mile Campground on left (Moenkopi Road)
38	19.6	N36 08.930 W115 22.500	Leaving Red Rock Canyon National Conservation Area

Trip 1 Continued

GPS Point	Cumulative mileage	Position	Comments
		Lat/Long ddd°mm.mmm'	Datum: WGS 84
39	20.6	N36 09.410 W115 21.697	Red Rock Basin flood detention basin on right
40	21.0	N36 09.533 W115 21.478	State Route 159 (Blue Diamond Rd.) becomes State Route 159 (Charleston Blvd.)
41	22.0	N36 09.528 W115 20.235	Intersection Charleston Blvd./I-215, end of trip 1

Trip 2

GPS Point	Cumulative mileage	Position	Comments
		Lat/Long ddd°mm.mmm'	Datum: WGS 84
42	0.0	N36 19.653 W115 18.730	Intersection, U.S. 95 and State Route 157, Kyle Canyon Road
43	8.6	N36 16.820 W115 27.057	Road to White Beauty Mine on left (East end, Harris Springs Rd.)
44	9.8	N36 16.727 W115 28.068	Kyle Canyon fan gravel/limestone contact on right
45	10.8	N36 16.652 W115 29.115	Gravel spires and blocks
46	12.3	N36 16.572 W115 30.877	Entering Spring Mountains National Recreation Area
47	12.6	N36 16.465 W115 31.247	Harris Springs Road (west end)
48	15.6	N36 16.228 W115 34.162	View of Charleston Peak
49	17.0	N36 16.182 W115 35.675	Entrance, Charleston Hotel parking lot
50	17.6	N36 15.858 W115 36.165	Lee Canyon turnoff (State Route 158) on right
51	18.0	N36 15.793 W115 36.572	Kyle Canyon Campground entrance on left
52	18.2	N36 15.790 W115 36.817	Entrance to Spring Mountains National Recreation Area Visitor Center on left
53	18.4	N36 15.775 W115 36.983	Fletcher View Campground on right
54	19.0	N36 15.683 W115 37.653	"Rainbow"
55	20.0	N36 15.583 W115 38.660	"Old Town"
56	20.9	N36 15.457 W115 38.970	Trailhead parking area
57	21.0	N36 15.422 W115 38.857	"Y" road intersection; End of road, retrace route to intersection with State Route 158
	0.0	*(odometer reset)*	
58		N36 15.858 W115 36.165	Intersection State Route 157/State Route 158; Set odometer to 0.0, turn left onto State Route 158
59	3.4	N36 18.165 W115 36.597	Robbers Roost information sign on right
60	4.5	N36 18.515 W115 36.430	Hilltop Campground turnoff on right
61	4.8	N36 18.537 W115 36.705	Trailhead, North Loop Trail
62	5.3	N36 18.730 W115 37.102	Mahogany Grove Group Camping Area on right
63	5.5	N36 18.853 W115 37.293	Deer Creek Picnic Area on right
64	6.3	N36 19.445 W115 37.313	Paleozoic limestone in roadcut
65	7.3	N36 20.185 W115 37.738	Desert viewpoint
66	8.9	N36 20.423 W115 39.137	Intersection with State Route 156, turn left onto State Route 156

GPS Coordinates for Trips

Trip 2 Continued

GPS Point	Cumulative mileage	Position	Comments
		Lat/Long ddd°mm.mmm'	Datum: WGS 84
67	10.7	N36 19.208 W115 40.340	Old Mill Picnic Area
68	10.9	N36 19.095 W115 40.427	Lee Canyon meadow
69	11.5	N36 18.645 W115 40.652	McWilliams Campground on right
70	11.7	N36 18.515 W115 40.738	Dolomite Campground on right
71	12.3	N36 18.397 W115 40.683	Turnaround at Bristlecone Trailhead; End of road, retrace route to intersection with State Route 158
	0.0	*(odometer reset)*	
72		N36 20.423 W115 39.137	Set odometer to 0.0, continue ahead on State Route 156
73	1.2	N36 21.293 W115 38.378	Sawmill Trailhead and picnic area on left
74	8.7	N36 25.357 W115 32.365	Leave Spring Mountains National Recreation Area
75	14.2	N36 28.615 W115 28.107	Intersection, State Route 156 and U.S. 95; end of trip, turn right to Las Vegas

Trip 3

GPS Point	Cumulative mileage	Position	Comments
		Lat/Long ddd°mm.mmm'	Datum: WGS 84
76	0.0	N36 11.743 W115 01.565	Starting point, intersection of Lake Mead Blvd. and Hollywood Blvd.
77	1.0	N36 11.922 W115 00.498	Great Unconformity
78	1.5	N36 11.973 W114 59.980	Kaibab Formation to left
79	2.3	N36 12.163 W114 59.183	Natural arch to left, skyline
80	3.8	N36 12.375 W114 57.710	Aztec Sandstone outcrop
81	5.0	N36 12.197 W114 56.472	PABCO road to left
82	5.6	N36 11.882 W114 55.902	Thumb breccia hill to right
83	5.8	N36 11.755 W114 55.675	view of Lava Butte at 2:30
84	6.3	N36 11.563 W114 55.318	Aztec/Horse Spring unconformity
85	6.8	N36 10.063 W114 54.668	More Thumb breccia
86	8.1	N36 09.660 W114 54.312	Large sign, pullout on right; entrance to Lake Mead National Recreation Area
87	8.7	N36 09.160 W114 54.312	Lake Mead National Recreation Area fee collection booth
88	9.3	N36 09.328 W114 53.770	Panoramic view to right
89	10.6	N36 08.588 W114 53.330	Intersection State Route 147 and Northshore Road, turn right onto Northshore Road
90	11.4	N36 08.740 W114 52.603	Paved turnout on right, NPS milepost 4
91	11.7	N36 08.718 W114 52.305	Road to upper Gypsum Wash on the right
92	12.0	N36 08.680 W114 51.932	Gypsum Wash
93	12.4	N36 08.702 W114 51.633	Paved turnout on right
94	13.1	N36 08.855 W114 50.938	Government Wash turnout on right
95	14.0	N36 09.178 W114 50.027	Photo turnout on right

Trip 3 Continued

GPS Point	Cumulative mileage	Position	Comments
		Lat/Long ddd°mm.mmm'	Datum: WGS 84
96	15.3	N36 09.473 W114 48.740	Paved turnout on left
97	15.6	N36 09.437 W114 48.392	Paved turnout on right
98	16.3	N36 09.415 W114 47.627	Turnoff to Boxcar Cove on right
99	17.4	N36 10.027 W114 46.905	Large, paved turnout on right
100	18.5	N36 10.653 W114 46.010	Callville Bay turnoff to right
101	20.1	N36 11.193 W114 44.537	Paved turnout to right ("road scars" turnout)
102	21.0	N36 11.598 W114 43.790	Paved turnout on left
103	21.9	N36 11.670 W114 42.810	Paved turnout on right
104	22.5	N36 11.775 W114 42.197	Paved turnout on right
105	23.0	N36 11.788 W114 41.587	Lovell Wash
106	23.3	N36 11.770 W114 41.307	Callville Wash
107	24.3	N36 12.082 W114 40.283	Syncline to left
108	25.5	N36 12.655 W114 39.390	Paved turnout on left
109	26.2	N36 12.790 W114 38.615	Paved turnout on left with Bowl of Fire interpretative sign
110	27.9	N36 13.608 W114 37.245	Northshore Summit Trailhead parking on left
111	29.2	N36 13.710 W114 35.995	Paved turnout on right
112	30.6	N36 13.798 W114 34.627	Paved turnout on left, view of Bitter Spring Valley
113	32.3	N36 14.523 W114 33.035	Paved turnout on left, this one with an emergency telephone
114	34.3	N36 14.560 W114 30.980	Redstone Rest Area
115	37.3	N36 15.842 W114 28.942	Boathouse Cove turn on the right
116	40.7	N36 18.540 W114 29.347	Echo Wash
117	42.7	N36 20.158 W114 29.465	Echo Bay Road to the right
118	44.6	N36 21.698 W114 28.887	Paved turnout
119	45.3	N36 21.777 W114 28.123	Wide shoulder on right; view of South Virgin Mountains
120	47.3	N36 22.647 W114 26.470	Rogers Spring Picnic Area to left
121	48.4	N36 23.352 W114 25.750	Blue Point Spring to left
122	48.8	N36 23.670 W114 25.502	Stewarts Point turn to right
123	49.8	N36 24.313 W114 25.103	Valley of Fire Wash
124	50.4	N36 24.660 W114 24.810	Salt Cove/Fire Cove Road
125	52.6	N36 26.235 W114 24.517	Overton Beach turnoff on right
126	54.0	N36 26.450 W114 25.643	Intersection, Northshore Road/Valley of Fire Road (to left)
	0.0	*(odometer reset)*	
127		N36 26.450 W114 25.643	Intersection, Northshore Road/Valley of Fire Road; reset odometer and continue ahead on Northshore Road
128	1.5	N36 27.668 W114 26.248	Leave Lake Mead National Recreation Area (Northshore Road becomes State Route 169)
129	5.8	N36 30.433 W114 25.648	Pistachio orchard on right
130	6.6	N36 30.992 W114 25.493	Intersection, East Waterfowl road to the right
131	6.7	N36 31.082 W114 25.582	Historical marker No. 41, Pueblo Grande de Nevada
132	7.0	N36 31.247 W114 25.747	Magnesite Road to the left

GPS Coordinates for Trips

Trip 3 Continued

GPS Point	Cumulative mileage	Position	Comments
		Lat/Long ddd°mm.mmm'	Datum: WGS 84
133	7.9	N36 31.885 W114 26.373	Entrance to Lost City Museum on left
134	8.6	N36 32.432 W114 26.587	End of sidetrip, Overton Park one block to right; retrace route to Valley of Fire Road
	0.0	*(odometer reset)*	
135		N36 26.450 W114 25.643	Intersection, Northshore Road/Valley of Fire Road; reset odometer to 0.0, turn right onto Valley of Fire Road
136	1.1	N36 25.907 W114 26.907	Enter Valley of Fire State Park
137	2.0	N36 25.745 W114 27.473	Elephant Rock, fee booth
138	2.3	N36 25.585 W114 27.773	Old Arrowhead Trail historical marker on left
139	2.6	N36 25.570 W114 28.098	John H. Clark historical marker in wash on left
140	3.1	N36 25.610 W114 28.590	Petrified log parking area on right
141	3.4	N36 25.610 W114 28.930	The Cabins Picnic Area to right
142	4.6	N36 25.577 W114 30.070	Seven Sisters Picnic Area
143	5.3	N36 25.592 W114 30.813	Turn right to Visitor Center and Mouse's Tank
144	5.4	N36 25.705 W114 30.813	Turn left to Mouse's Tank (or continue ahead to Visitor Center)
145	6.4	N36 26.457 W114 30.988	Mouse's Tank parking area; turn around and return to Valley of Fire Road
	0.0	*(odometer reset)*	
146		N36 25.592 W114 30.813	Intersection, Valley of Fire Road; reset odometer to 0.0, turn right and continue through park
147	1.9	N36 25.297 W114 32.645	Paved road to Atlatl Rock and campground, and Arch Rock Campground on right"
148	2.6	N36 24.820 W114 32.902	The Beehives
149	3.7	N36 24.347 W114 34.007	State Park exit station
150	5.3	N36 24.667 W114 35.145	Top of old alluvial fan
151	7.4	N36 24.967 W114 37.055	View of limestone on Piute Point
152	18.2	N36 30.118 W114 45.678	Intersection with I-15, pass under freeway, turn left and return to Las Vegas
153	29.5	N36 22.863 W114 53.538	Exit for U.S. 93, "Great Basin Highway," to right

Trip 4

GPS Point	Cumulative mileage	Position	Comments
		Lat/Long ddd°mm.mmm'	Datum: WGS 84
154	0.0	N36 02.393 W114 58.918	Intersection, Boulder Highway (State Route 582) and Lake Mead Parkway (State Route 564)
155	2.7	N36 04.337 W114 57.337	Intersection, Athens and Lake Mead Parkway ("white rock")
156	3.2	N36 04.542 W114 56.913	Intersection, Calico Ridge Drive and Lake Mead Parkway

Trip 4 Continued

GPS Point	Cumulative mileage	Position	Comments
		Lat/Long ddd°mm.mmm'	Datum: WGS 84
157	3.7	N36 04.637 W114 56.372	Golda Way, River Mountains Aqueduct
158	4.7	N36 05.040 W114 55.417	Lake Las Vegas Parkway on left
159	5.3	N36 05.268 W114 54.758	Former entrance, Three Kids Mine on right
160	6.1	N36 05.853 W114 54.265	State Historical Marker 141 on right
161	6.2	N36 05.957 W114 54.230	Intersection, Lorin L. Williams Parkway, to left
162	6.3	N36 06.110 W114 54.148	Enter Lake Mead National Recreation Area (Lake Mead Parkway now becomes Lakeshore Road)
163	6.5	N36 06.212 W114 54.103	Paved turn around on right, tortoise fence
164	6.7	N36 06.323 W114 54.062	Fee collection booth, Lake Mead National Recreation Area
165	7.1	N36 06.588 W114 53.973	Junction, Lakeshore Road and Northshore Road
	0.0	*(odometer reset)*	
166		N36 06.588 W114 53.973	Intersection, Lakeshore Road/Northshore Road; reset odometer to 0.0, turn left onto Northshore Road
167	1.0	N36 07.377 W114 54.293	Parking area on north side of bridge; end of trip, return to Northshore Road/Lakeshore Road intersection
	0.0	*(odometer reset)*	
168		N36 06.588 W114 53.973	Intersection, Northshore Road/Lakeshore Road; Turn left onto Lakeshore Road, continue eastward
169	0.8	N36 07.072 W114 53.295	Wash, Muddy Creek Formation and turtle fences
170	1.8	N36 07.078 W114 52.300	Road to Las Vegas Bay
171	2.1	N30 06.940 W114 52.033	Las Vegas Bay overlook turn
172	3.5	N36 06.372 W114 50.735	Road to The Cliffs Overlook (33 hole picnic area)
173	4.1	N36 05.898 W114 50.630	Paved road to Long View scenic overlook
174	5.1	N36 05.732 W114 49.645	Desert View scenic overlook
175	5.9	N36 04.770 W114 49.448	Road to Lake Mead Fish Hatchery
176	6.0	N36 04.662 W114 49.373	Saddle Island Road
177	6.8	N36 03.982 W114 49.192	A.M. Smith Water Treatment Facility
178	8.2	N36 02.835 W114 48.988	Lake Mead Marina entrance
179	8.7	N36 02.545 W114 48.573	Lake Mead Lodge Road
180	9.1	N36 02.213 W114 48.308	Boulder Beach
181	9.4	N36 02.070 W114 48.198	Boulder Beach Campground
182	10.2	N36 01.378 W114 47.828	Hemenway Harbor access
183	10.4	N36 01.240 W114 47.767	Lake Mead National Recreation Area fee collection booth
184	11.0	N36 00.730 W114 47.657	Turn to Trailhead parking for Historic Railroad Trail

GPS Coordinates for Trips

Trip 4 Continued

GPS Point	Cumulative mileage	Position	Comments
		Lat/Long ddd°mm.mmm'	Datum: WGS 84
185	11.1	N36 00.660 W114 47.748	Turn to Alan Bible Visitor Center, to left
186	11.4	N36 00.550 W114 47.998	Intersection, Lakeshore Road/U.S. 93
	0.0	*(odometer reset)*	
187		N36 00.550 W114 47.998	Intersection, Lakeshore Road/U.S. 93, reset odometer to 0.0, turn left onto U.S. 93
188	0.9	N36 00.567 W114 47.035	Hacienda Hotel and Casino on the left
189	1.2	N36 00.645 W114 46.657	Intersection, new U.S.93 bypass on right (approximate location, under construction)
190	2.6	N36 00.872 W114 45.603	Turn off to Lakeview Overlook on left
191	2.7	N36 00.892 W114 45.438	Lower Portal Road to the right
192	2.9	N36 00.840 W114 45.282	Security checkpoint
193	4.2	N36 00.917 W114 44.463	Entrance to covered parking on left
194	4.4	N36 00.985 W114 44.230	Nevada/Arizona State Line, center of Dam
195	4.5	N36 00.900 W114 44.208	Fault "sign" ahead; continue ahead to any safe turnaround, then retrace route back across dam
	0.0	*(odometer reset)*	
196		N36 00.997 W114 44.342	At Winged Victory statue, Nevada end of Dam, continue west to Boulder City
197	5.5	N36 00.237 W114 49.087	"Welcome to Boulder City" sign
198	5.7	N36 00.033 W114 49.257	Boulder City, city limit sign
199	5.9	N35 59.873 W114 49.390	Pacifica Way, River Mountains Trailhead parking to right
200	6.5	N35 59.412 W114 49.837	Intersection; business 93 through Boulder City to left, 93 bypass straight ahead
201	7.2	N35 59.020 W114 50.560	Turnoff to River Mountains Hiking Trail trailhead parking and display
202	7.9	N35 58.458 W114 50.783	Western intersection, business 93 and 93 bypass, turn right toward Las Vegas
203	8.5	N35 58.187 W114 51.310	Yucca Drive on right, access to Bootleg Canyon Trails and Nevada State Railroad Museum
204	10.8	N35 58.080 W114 53.660	Junction with U.S. 95, exit to right, turn left under freeway, sidetrip to Nelson begins
	0.0	*(odometer reset)*	
205		N35 58.080 W114 53.660	At intersection, travel south on U.S. 95
206	2.0	N35 56.487 W114 54.445	Fault along base of McCullough Range to right
207	3.9	N35 55.205 W114 55.068	Crossing under power lines
208	10.0	N35 49.707 W114 56.235	Intersection, U.S. 95 and SR 165; turn left (east) onto SR 165 to Nelson
209	13.4	N35 48.472 W114 52.798	Powerlines overhead
210	17.5	N35 45.562 W114 50.910	Small quarry on left
211	20.5	N35 42.643 W114 49.615	Entering Nelson
212	23.0	N35 42.575 W114 48.223	Techatticup Mine and junk on right, small store on left
213	23.5	N35 42.468 W114 47.683	Stopes ahead on ridge
214	24.0	N35 42.433 W114 47.110	Eldorado fault

Trip 4 Continued

GPS Point	Cumulative mileage	Position	Comments
		Lat/Long ddd°mm.mmm'	Datum: WGS 84
215	24.4	N35 42.582 W114 46.630	Entering Lake Mead National Recreation Area (sign)
216	26.0	N35 42.673 W114 45.150	Hoodoos on right
217	28.8	N35 42.512 W114 42.815	Paved road down into wash and former site of Nelsons Landing, to right
218	29.0	N35 42.557 W114 42.650	Loop road at end of State Route 165; end of sidetrip, retrace route to U.S. 95/U.S. 95 intersection
	0.0	*(odometer reset)*	
219		N35 58.080 W114 53.660	Intersection, U.S. 95/U.S. 93, reset odometer to 0.0, turn left under freeway and continue to Henderson
220	1.1	N35 58.295 W114 54.698	Railroad Pass Hotel and Casino on right
221	1.5	N35 58.732 W114 54.920	Railroad Crossing
222	3.1	N35 59.737 W114 56.055	Exit 56B to Boulder Highway from I-515
223	4.3	N36 00.507 W114 56.795	Clark County Heritage Museum on right
224	7.2	N36 02.393 W114 58.918	Intersection, Boulder Highway (State Route 582) and Lake Mead Parkway (State Route 564); end of trip

Trip 5

GPS Point	Location	Position	Comments
		Lat/Long ddd°mm.mmm'	Datum: WGS 84
225	U.S. 95 North/Corn Creek Rd.	N36 25.553 W115 25.400	Intersection is about 9.3 miles north of Kyle Canyon Road intersection
226	Corn Creek Springs	N36 26.308 W115 21.495	Corn Creek Springs Visitor Center, about 4 miles of gravel road east of U.S. 95
227	Tule Springs	N36 19.053 W115 16.275	Entrance to Floyd Lamb State Park at Tule Springs
228	Kyle (Kiel) Ranch Spring	N36 12.200 W115 08.428	Nevada Historical Marker No. 224
229	Las Vegas Springs	N36 10.217 W115 11.538	Entrance to the Springs Preserve
230	U.S. Post Office Benchmark	N36 10.343 W115 08.485	Benchmark located to right of front steps, historical Las Vegas Post Office
231	Valley View Blvd. at Charleston	N36 09.638 W115 11.563	On "up" side of fault, on Valley View north of Charleston
232	Cheyenne Ave. and Revere St.	N36 13.083 W115 08.870	Near bottom of scarp, Windsor Park/Cashman Field fault zone
233	Sunset Rd. west of Arroyo Gr.	N36 03.878 W115 03.840	Whitney Mesa fault zone
234	Henderson Bird Viewing Preserve	N36 04.338 W115 00.097	Entrance to facility
235	Clark County Wetlands Park	N36 06.057 W115 01.357	Parking area in front of Visitor Center